# 4. Position and Shape

**THE OPEN UNIVERSITY**

Technology/Mathematics: A Second Level Course

**MODELLING
BY MATHEMATICS**
TM 281

# Block 2
# POSITION AND MOVEMENT

# 4. Position and Shape
# 5. Lines, Curves and Directions
# 6. Movement

*Prepared by the Course Team*

THE OPEN UNIVERSITY PRESS

## THE MODELLING BY MATHEMATICS COURSE TEAM

| | |
|---|---|
| David Blackburn | (Chairman) |
| Phil Ashby | (BBC) |
| Keith Attenborough | (Technology) |
| Gerald Copp | (Editor) |
| Peter Cox | (Student Computing Service) |
| Bob Davies | (Senior Counsellor) |
| Judy Ekins | (Mathematics) |
| Roger Harrison | (Institute of Educational Technology) |
| Don Hurtley | (Staff Tutor) |
| Maurice Inman | (Staff Tutor) |
| David Johnson | (Student Computing Service) |
| Roy Knight | (Mathematics) |
| Ernest Law | (Staff Tutor) |
| Owen Lawrence | (Staff Tutor) |
| Daniel Lunn | (Mathematics) |
| Patricia McCurry | (BBC) |
| Alistair Morgan | (Institute of Educational Technology) |
| Colin Robinson | (BBC) |
| John Sparkes | (Technology) |
| Bob Tunnicliffe | (Mathematics) |
| Mirabelle Walker | (Technology) |
| Geoffrey Wexler | (Technology) |

The Open University Press
Walton Hall, Milton Keynes
MK7 6AA

First published 1977. Reprinted with corrections 1982.

Copyright © 1977 The Open University

Designed by the Media Development Group of the Open University.

Produced in Great Britain by
Technical Filmsetters Europe Limited, 76 Great Bridgewater Street, Manchester M1 5JY

ISBN 0 335 06291 1

This text forms part of an Open University course. The complete list of units in the course appears at the end of this text.

For general availability of supporting material referred to in this text, please write to Open University Educational Enterprises Ltd, 12 Cofferidge Close, Stony Stratford, Milton Keynes, MK11 1BY.

Further information on Open University courses may be obtained from the Admissions Office, The Open University, P.O. Box 48, Walton Hall, Milton Keynes, MK7 6AB.

# CONTENTS

## AIMS

The aims of this unit are:

1   To introduce you to the vocabulary of geometry.

2   To show how geometrical figures can be used as models for solving real problems, in particular those dealing with the measurement of height and area, and the construction of right-angles.

3   To demonstrate the validity of these geometrical procedures by explaining the properties of parallel lines and triangles.

4   To introduce some of the geometrical properties of circles.

5   To introduce you to trigonometry and illustrate its applications.

## OBJECTIVES

When you have finished this unit you should be able to:

1   Distinguish between true and false statements concerning, or explain in your own words the meaning of, the following terms:

| | |
|---|---|
| acute angle | perpendicular |
| alternate angles | plane |
| arc of a circle | protractor |
| bisector | radian |
| chord of a circle | radius |
| circle | ratio |
| circumference of a circle | right-angle |
| congruent triangles | sector of a circle |
| converse | semi-circle |
| corresponding angles | similar triangles |
| degree | sphere |
| diameter | straight line |
| equilateral triangle | tangent to a circle |
| isosceles triangle | triangle |
| line | vertex of a triangle |
| obtuse angle | vertically opposite angles |
| parallel lines | |
| parallelogram | |

2   Recognize angles that are equal in situations involving intersecting, parallel and perpendicular lines and use these equalities to show that the angles of a triangle add up to 180° and to establish when triangles are congruent or similar (SAQs 1, 3, 4, 5, 8, 15, 16).

3   List the sets of information that show that triangles are congruent and construct a triangle given any one of these sets of information (SAQ 2).

4   State what is meant by similar triangles and list their main properties. Use these properties in problems involving similar triangles, such as the shadow pole (SAQs 9, 10, 11, 12, 13).

5   Calculate the area of a triangle, a parallelogram, a circle and the surface of a cylinder. Derive the formula for area in the case of the triangle and parallelogram (SAQs 4, 5, 6, 7).

6   State Pythagoras' theorem. Use similar triangles to prove it and use its converse for the construction of right-angles. Use the theorem to calculate the length of a side in a right-angled triangle, given the other two sides (SAQs 13, 14).

7   Bisect an angle and construct right-angles using a compass and rule (SAQ 15).

8   Define sine, cosine and tangent.

9 Use tables of sine, cosine and tangent (SAQs 19, 20, 23, 28, 31, 32).

10 Use the sine, cosine or tangent relationship, as appropriate, to calculate the other sides and angles in a right-angled triangle (SAQs 24, 25, 26, 27, 28).

11 Sketch graphs of sine, cosine and tangent so as to show their symmetries and periodicities (SAQ 30).

12 Describe how to do a simple survey and make calculations of the type needed in such surveys (SAQs 33, 34).

13 Find the other sides and angles in any given triangle, where either all three sides, two sides and an angle or one side and two angles are known (SAQs 34, 35).

## STUDY GUIDE

Your work for this study week consists of Unit 4, *Position and shape*, the first side of Disc 3, *Using trigonometric tables*, and TV3, *Nothing New Under the Sun*. The assignment material associated with this unit is printed in the supplementary material.

You are expected to be familiar with the material in Block 1; graphs and the manipulation of algebraic symbols and expressions, before starting this unit.

This unit aims to teach some geometry and trigonometry. It is possible that you may have met some of the material before, but I do not advise you to skip any sections. *You are not expected to remember the formulae and results given in this unit*, but you are expected to be able to use them. The important results are given in your *Handbook* and you should get used to referring to it while working on the self-assessment questions and assignments.

The unit teaches you how to use trigonometric tables and Disc 3, Side 1, takes you through some examples of their use. If you are already familiar with the use of tables and can obtain the correct answers to the relevant self-assessment questions, do not bother to listen to the disc. If, however, you are not too confident about the use of these tables, I advise you to make good use of the disc. You may find it useful to stop the disc and play it over again as it covers quite a lot of material.

It is important that you become familiar with the use of trigonometric tables, although for specific problems you will often find that your slide rule gives the trigonometric functions with sufficient precision for your purpose. Your *Slide Rule Book* explains how this is done. It is not, however, essential that you learn to use the slide rule for this purpose as you can always use the more accurate tables. If you have a calculator, this may also have a facility for finding the values for trigonometric functions.

The television programme, TV3, *Nothing New Under the Sun*, uses some geometric ideas and models to explain the construction of a sundial. The function of this programme is to let you see geometry in action, so do not worry if you are unable to follow all the arguments at the speed they are presented. You will, however, benefit more from the programme if you have read Sections 1–5 of this unit before viewing: you will find more on the sundial in Section 5.6. Note that the three dimensional aspects of this section are not considered as examinable material for this course.

When you have finished the unit you can use the objectives and summary in conjunction with your *Handbook* (the geometry and trigonometry sections) as a checklist of what you are expected to know and do as a result of studying the unit.

# 1 INTRODUCTION

Units 4, 5 and 6 are concerned with the ways in which we describe position and movement. Today, this is something largely taken for granted. We have maps and globes; and road signs labelled with distances that we assume to be correct. We travel by train, ship or aeroplane and assume that we shall arrive at our chosen destination more or less at the advertised time. Areas of land are calculated correctly, buildings are set out with right-angles at their corners and are built with vertical sides. Buried in all these activities are various kinds of mathematical models.

Units 4 and 5 deal with the position, size and shape of things and so the mathematical models used to represent reality are basically geometrical. This unit concentrates on triangles, squares, rectangles and circles. Unit 5 takes up the circle again and discusses other curves that are useful in forming mathematical models. Geometrical shapes such as triangles or squares, or even lines, are called models because, in geometry, the lines are all straight or follow smooth curves and they are thought of as having no thickness. Similarly, triangles and squares are thought of as flat with no bumps in them. In reality, however, nothing is quite like that. No lines are quite straight; many things that are represented by lines, such as poles or lengths of string, or even the line we draw on a piece of paper to represent a 'straight line', have some thickness. Plots of land or floors are not quite flat and are not simple geometrical shapes. The geometrical ideas we use are therefore models of reality; but by studying the properties of these models we can perform useful calculations about the world around us and can do things we would be unable to do without the insight they give.

I do not expect you to know much geometry as you begin this unit and so some of the following may be too elementary for you. However, a good many words I shall use—such as vertical, horizontal, circle, square—have their origins in geometry, even though they are now used in everyday language. I shall therefore build upon what I think it is reasonable to expect you to know about them or to find out from a dictionary.

## 2 MEASURING THE HEIGHT OF A TREE

Suppose you have a tall tree in your garden that you would like to cut down. You wonder whether it would fall on your house or into your neighbour's garden. To answer this question you need to know the height of the tree. How can you measure it without actually climbing up the tree and using a long tape measure?

One method would be to set up a vertical stick of known height (say two metres), measure the length of its shadow and compare this with the length of the tree's shadow. If the tree's shadow is 10 m long when the pole's shadow is 1 m long, you might conclude that the tree was ten times as long as the stick—twenty metres. One of the questions I want to tackle is how this method, which in many situations will give a good estimate of the tree's height, can be justified.

During the summer, when the Sun is high in the sky, another simple way might be to wait until the length of the stick's shadow is the same as that of the stick, 2 m. Then it would be reasonable to conclude that the height of the tree would be equal to the length of its shadow. This is illustrated in Figure 1. Measure the length of the shadow and you have the length of the tree.

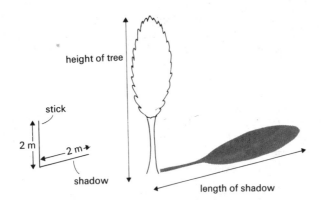

*Figure 1   Estimating the height of a tree by measuring its shadow.*

In order to estimate how the tree might fall you could get a length of string and cut it to be the length of the shadow. You could then use the string to mark out a circle with the base of the tree as its centre and the height as its radius. Where the circle and the radius lie over clear ground should be a safe place to fell the tree.

Figure 2 shows two geometrical models of the practical task of measuring the height of a tree, using the method illustrated in Figure 1, of waiting until

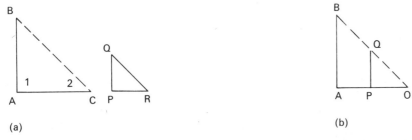

(a)                                                                 (b)

*Figure 2   A geometrical model of the problem of Figure 1. (a) In the two triangles AB represents the tree, PQ represents the stick and AC and PR represent their shadows. (b) Shows the triangles placed on top of each other so that they have a common angle at O. The shadows are now represented by AO and PO.*

the stick's shadow equals its height. Let me list some of the conventions used in this model. They are the normal conventions of geometry and will be used in this unit and throughout the course. They are part of the business of setting up mathematical models.

1   The tree, the stick, their shadows and the supposed path of a ray of sunlight from the top of the tree or stick to its shadow are represented by *lines*. As I have already remarked, they are to be thought of as having no thickness, although they are drawn with a small but measurable thickness. In this case, the lines are straight lines because this seems the best way to represent the objects in question. Actually, trees may lean or be curved and the ground on which the shadows fall may be uneven. In such cases, the model will be less precise than if the ground were flat.

line

2   The lines are labelled by letters placed at the ends of the lines. In Figure 2(a), AB represents the tree and AC represents its shadow. AO represents the tree's shadow in Figure 2(b). PQ represents the two-metre stick.

What do BC, PR, OQ represent in Figure 2?

BC represents the path of the ray of sunlight passing from the top of the tree to the ground. PR represents the pole's shadow and OQ represents the path of the ray of sunlight passing from the top of the pole to the ground.

3   If two lines meet, they meet at a *point*. Equally, any line ends at a point and therefore the letters used to label lines also label points. I may thus speak of line AB ending at points A and B. Points are thought of as having no size at all.

4   A triangle is identified or named by the three letters placed at each corner or *vertex* of the triangle.

vertex

In Figure 2 there are four triangles. The names of two of them are ABC and OPQ. What are the names of the other two?

They are PQR and AOB. The letters can be listed in any order, so QRP, RPQ, PRQ, etc., are all correct names for the small triangle in Figure 2(a).

5   If I want to refer to an *angle*, I normally use three letters to name it. The first and last letters name the remote ends of the lines which enclose the angle. The letter which identifies the point where the angle occurs is placed at the centre of the three letters and has a circumflex over it. I have marked two angles in Figure 2(a) with numbers. The angle labelled (1) is normally named $\hat{BAC}$ or $\hat{CAB}$.

How would you name the angle marked (2)?

$\hat{BCA}$ or $\hat{ACB}$

If there is no danger of ambiguity you can use a single letter to identify an angle. Thus, $\hat{BAC}$ can be labelled $\hat{A}$; but $\hat{OPQ}$ in Figure 2(b) cannot be labelled $\hat{P}$ because the angle $\hat{APQ}$ is also at point P.

Let us return to the particular model illustrated in Figure 2 and consider what particular features of these diagrams make them applicable to the task illustrated in Figure 1.

Because the ground is supposed to be horizontal and the tree and stick are supposed to be vertical, BA and AC, and QP and PR are drawn *perpendicular* to each other; angles $\hat{BAC}$ and $\hat{QPR}$ are called *right-angles*. It will be assumed that you are familiar with the idea of measuring angles in *degrees* and that you can use a *protractor* to do this. Figure 3 shows a protractor: using it you could construct lines that are *perpendicular* to each other, that is, lines that are at *right-angles*, with an angle of 90° between them. If four right-angles are arranged as in Figure 4, they go right round the point X. The angle around a point is therefore 360°. The angle at a point on a straight line is 180° or two right-angles.

perpendicular
right-angle
degree
protractor

Figure 3 *A protractor is used for measuring angles.*

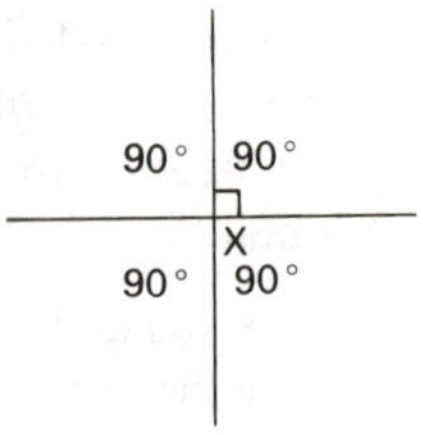

90° | 90°
90° | 90°

Figure 4 *Four right-angles arranged around a point add up to 360°. Two right-angles stand side by side on a straight line and make an angle of 180°. Note the symbol for a right-angle.*

For the given position of the Sun, the height of the stick is the same as the length of its shadow and therefore the triangle QPR is drawn so that $QP = PR$. The method used implies that, similarly, $AB = AC$ and that therefore the height of the tree is also equal to the length of its shadow. In Section 5.3, I shall show why this conclusion is valid. A triangle with two sides equal in length is called an *isosceles triangle*. Triangles ABC and QPR are, therefore, both isosceles triangles. Isosceles triangles do not necessarily have a right-angle between the equal sides. Figure 5 shows several examples

**isosceles triangle**

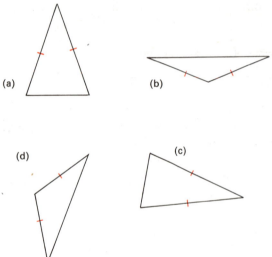

Figure 5 *Four differently shaped isosceles triangles. The lines that are marked in red are equal.*

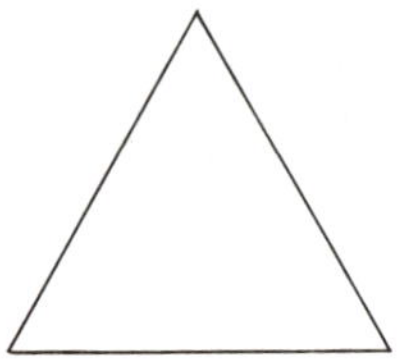

Figure 6 *This triangle has all three sides of equal length and is called an equilateral triangle.*

of isosceles triangles with different angles, but they all have a pair of equal sides. If three sides of a triangle are equal, it is called an *equiliteral triangle* (Figure 6). An angle which is less than a right-angle is called an *acute angle*; one which is greater than a right-angle, but less than two right-angles, is called an *obtuse angle*.

**equilateral triangle**

**acute and obtuse angles**

Which of the isosceles triangles in Figure 5 have acute angles between their equal sides?

Triangles (a) and (c)

The final point about Figure 2 which makes the method of measurement work is in fact that rays of sunlight are very nearly parallel to each other. As a consequence the two sides of triangles ABC and PQR which represent the paths of the Sun's rays (i.e. lines BC and QR) are drawn parallel to each other, just as are the two vertical lines representing the stick and the tree.

*Parallel lines* are lines that never meet, however far they are extended. Because the rays may have started at the same point on the Sun, they are not exactly parallel. To model them with parallel lines is, however, a close approximation.

The argument implicit in the method used to measure the height of the tree illustrated in Figure 2 is as follows:

> 'If we hold a stick vertical so that it is parallel with the tree, and if the ground on which the shadows fall is horizontal, and if the Sun's rays are parallel, then the triangles formed by the lines representing the tree and its shadow, the stick and its shadow, and the Sun's rays are *similar* in the sense that if two sides of one triangle are equal then two sides of the other triangle are also equal.'

One of the tasks of geometry is to prove that inferences like these are valid. In this course, Euclidean geometry, based on the system set out by Euclid in the third and second centuries BC, is used. There are other kinds of geometry (non-Euclidean geometry), but I shall not be concerned with them here. I shall not work through all the theorems, but shall pick out those arguments and ideas which are most widely used today as mathematical models for practical purposes.

The general geometrical problem which has emerged from this shadow-pole problem is the question of what constitutes similar triangles and what inferences you can draw about one triangle from knowledge about the other. This and similar questions are ones to which I shall turn shortly, but first I must lay some basic foundations of the subject.

# 3 SOME GEOMETRICAL IDEAS

In this course I shall only be concerned with shapes and lines on flat surfaces. Shapes and lines on curved surfaces, such as the surface of a sphere or cylinder, have quite different properties.

## 3.1 Parallel and intersecting lines

You know what straight and flat mean, but can you express these ideas in terms of concise definitions? A *straight line* is the shortest distance between two points. A flat surface, called a *plane*, is one on which a straight line can be drawn in any direction so that it lies completely within the plane. This obviously does not refer to a straight line which sticks out of the plane and intersects it at only one point.

**straight line**

**plane**

Two straight lines on a flat surface either meet if they are extended far enough, or else they are parallel lines. (If the lines are not confined to the same plane they may never meet and yet be non-parallel; such lines are called 'skew' lines and they are discussed in books on solid geometry.)

Two straight lines that intersect, as in Figure 7, form angles with each other at the point of intersection. The lines AB and CD intersect at X and form four angles: $A\hat{X}D$, $A\hat{X}C$, $B\hat{X}D$ and $B\hat{X}C$. Some of these angles are the same size. We have already noted that the angles on a straight line add up to 180°. (Two right-angles, back to back so to speak, form a straight line—see Figure 4.) This means that:

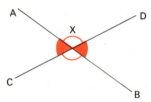

$$A\hat{X}D + B\hat{X}D = 180°$$

because they are angles on the straight line AB; and

$$A\hat{X}D + A\hat{X}C = 180°$$

*Figure 7  Two intersecting straight lines AB and CD. Vertically opposite angles are equal.*

because they are angles on the straight line CD.

Now subtract one equation from the other

$$A\hat{X}D + B\hat{X}D = 180°$$
$$A\hat{X}D + A\hat{X}C = 180°$$
$$\overline{B\hat{X}D - A\hat{X}C = \quad 0°}$$

Therefore $B\hat{X}D = A\hat{X}C$.

These two angles are called *vertically opposite angles* and are shaded in Figure 7. Vertically opposite angles are always equal to each other. The unshaded angles, $A\hat{X}D$ and $B\hat{X}C$, are also vertically opposite and are also equal to each other.

**vertically opposite angles**

Under what circumstances are all four angles equal?

When the intersecting lines are at right-angles to each other.

I hope you can see that these comments are equally true whatever the angle of intersection of the lines AB and CD.

If $B\hat{X}D = 40°$ what size are the other three angles around the point X?

$A\hat{X}D = 140°$: $C\hat{X}A = 40°$: $B\hat{X}C = 140°$.

If a line intersects two parallel lines, as in Figure 8(a), several more angular equalities emerge. The angles $B\hat{O}Y$ and $C\hat{P}X$ are called *alternate angles* and are equal. In Figure 8(b), angles $X\hat{O}B$ and $D\hat{P}X$ are called *corresponding angles* and are equal. They will be discussed from a different standpoint in Section 5.1.

**alternate angles**

**corresponding angles**

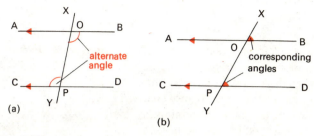

Figure 8   *Parallel lines intersected by a single line. (a) shows alternate angles and (b) shows corresponding angles. Parallel lines are indicated by arrows.*

**SAQ 1**

In Figure 9, AB and CD are parallel lines.

(a)   Identify which angles are alternate angles.

(b)   Identify which angles are vertically opposite angles.

(c)   Give two examples of a pair of angles that add up to 180°.

(d)   Give two examples of corresponding angles.

SAQ 1

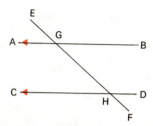

Figure 9   *See SAQ 1.*

These considerations of angular equalities and pairs of angles that add up to 180° lead to a very important result: that the *internal angles of a triangle add up to 180°*. You can show that this is true as follows.

Draw a triangle, any triangle, as in Figure 10(a). Extend the base of the triangle and draw a line through one of the vertices that is parallel to the opposite side, as shown in Figure 10(b). I have marked a pair of equal *corresponding angles x* and a pair of equal *alternate angles y* (Figure 10(c)). The internal angles of the triangle are $x$, $y$ and $z$; but $x$, $y$ and $z$ are also the angles on a straight line, so they add up to 180°. Therefore, *the angles of any triangle add up to 180°*. This proof is typical of a geometrical proof and of a general conclusion which is not obvious at first sight.

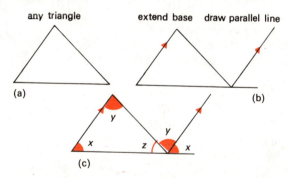

Figure 10   *Proof that the angles of a triangle add up to 180°. Angles marked x are corresponding angles and angles marked y are alternate angles. Angles x, y and z add up to 180°.*

If two parallel lines intersect two other parallel lines a *parallelogram* is formed as shown in Figure 11(a). Opposite sides of a parallelogram are equal in length. (See SAQ 3, Section 3.2.) If the pairs of parallel lines are at right-angles to each other the parallelogram is a *rectangle* (Figure 11(b)). A *square* is a rectangle with all sides equal.

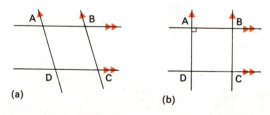

Figure 11   *(a) Two intersecting pairs of parallel lines form a parallelogram, ABCD. (b) If they are at right angles, ABCD is a rectangle.*

15

## 3.2 Congruent triangles

Triangles which are the same shape and size are called *congruent triangles*. If you were to cut congruent triangles out of a piece of paper, you would be able to fit one exactly on top of the other, though you might have to turn one over to do so (Figure 12). The general question, which Euclid answered long ago, and you will need to know the answer to, is what is the minimum information about two triangles that has to be specified in order to be sure that the two triangles are congruent?

congruent triangles

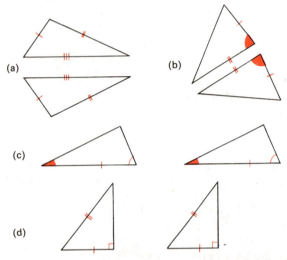

*Figure 12   Pairs of congruent triangles: (a) three sides equal; (b) two sides and the angle between them equal; (c) one side and the angles at its ends equal; (d) any two sides equal and a right-angle. For each pair of congruent triangles equal lines and equal angles are marked the same.*

Every triangle has three sides and three internal angles and in two congruent triangles all six features of one are equal to the corresponding six features of the other. However, to know that two triangles are congruent, you do not need to find out if all six features are equal. If you know two angles of a triangle you can always calculate the third because the sum of all three is 180°. Evidently, it is not necessary to check the equality of all three angles to check for congruence. On the other hand, even if you know that two sides are the same in two triangles you cannot assume that the other sides are equal. Figure 13 shows three triangles in which one side is 2 cm and another is 3 cm long; they are clearly not congruent triangles, since the remaining sides are all of different length.

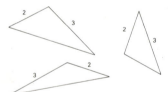

*Figure 13   Triangles in which two sides are specified (e.g. 2 cm and 3 cm). The triangles are not congruent, however, because the third sides are obviously not equal.*

What information is just sufficient to ensure congruence?

You will know that triangles are congruent if any one of the following conditions hold:

(a)      Three sides are equal (Figure 12(a)).

(b)      Two sides and the angle between them are equal (Figure 12(b)).

(c)      One side and the angles at each end of it are equal (Figure 12(c)).

(d)      Two sides are equal and one of the angles is a right-angle (Figure 12(d)).

You do not need to prove this.

Just knowing the equality of two or three angles is *insufficient*; this will be discussed in Section 5.2.

Another way of thinking about these sets of information is to regard each as a recipe for constructing a triangle without danger of ambiguity.

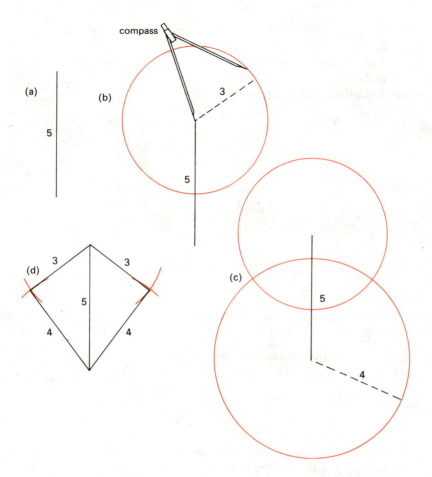

*Figure 14   Constructing a triangle whose sides are of length 3, 4 and 5 cm. (a) Draw a line 5 cm long. (b) Draw a circle, centred at one end of the 5 cm line, of radius 3 cm. (c) Draw a circle of radius 4 cm centred on the other end of the 5 cm line. (d) The third vertex is at an intersection of the circles.*

For example, to construct a triangle when you are given the length of all three sides you proceed as follows (Figure 14). First draw one of the sides, preferably the longest. Then place a compass point at one end of the line and draw a circle whose radius is equal to one of the other lengths. Now place the compass point at the other end of the line you first drew and draw a circle whose radius is equal to the third length. Where the circles intersect is where the third vertex of the triangle must be. There are two intersections and two possible solutions (Figure 14(d)), but one triangle is the mirror image of the other, so the two solutions are congruent—they could be cut out and fitted on top of each other exactly, though you would have to turn one over.

### SAQ 2

SAQ 2

Suppose you are provided with a ruler and protractor (you do not need a compass this time):

(a)  How would you draw a triangle given that two sides are each 10 cm long and the angle between them is 30°?

(b)  How would you draw a triangle given that one side is 10 cm and that the angles at each end of it are 30°?

### SAQ 3

SAQ 3

Prove that the opposite sides of a parallelogram are equal in length. Hint: You must create a pair of congruent triangles.

If you are given the lengths of two sides and an angle other than the one between them, it is possible that you will be able to draw two *different* (non-congruent) triangles each meeting the description given (see Figure 15). If the angle given is a right-angle, however, the danger of ambiguity does not arise.

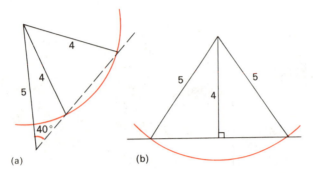

(a)                    (b)

*Figure 15   (a) You are given that a triangle has two sides of 4 cm and 5 cm, and that the angle at the end of the 5 cm line, not between it and the 4 cm line, is 40°. You can construct two possible triangles, but they are not congruent with each other. (b) If you are given two sides of a triangle and know that one of the angles is a right-angle, two triangles can again be drawn, but this time they are congruent.*

## 3.3   Circles

A circle is a very simple curved geometrical figure. It will be used as a model during the course in a number of situations. I shall consider some of its geometrical properties later in this unit and in Unit 5. Here, I am only concerned with the meaning of some of the words that apply to it.

A *circle* is a line such that every point on it is the same distance (called the *radius, r*) from a given point called its centre. The length of this curved line is called the *circumference*, it is $2\pi$ times the radius. Thus in Figure 16(a)

**circle**
**radius**
**circumference**

$$\text{circumference} = 2\pi r$$

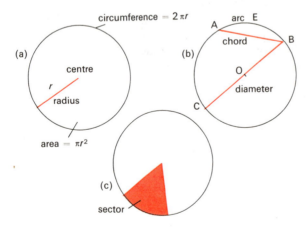

*Figure 16   Terminology of circles.*

The value of $\pi$ is 3.14 (to three significant figures).

The area enclosed by the circle is $\pi r^2$. (A proof of this is given later in the course.)

Figure 16(b) introduces some further terms. A straight line from one point on the circumference to another is called a *chord*. A chord which goes through the centre is a *diameter*. An *arc* of a circle is a portion of the circumference. Two arcs are delineated by the chord AB. The short arc is

**chord**
**diameter**
**arc**

18

labelled AEB to distinguish it from the much longer arc ACB. Figure 16(c) shows a *sector* of a circle. A sector is bounded by an arc and two radii. A *semi-circle* is bounded by a diameter and half of the circumference (e.g. CAEB of Figure (16(b)).

sector

semi-circle

I have already referred to degrees, by which angles are measured. They can be defined as follows. Suppose the line OC in Figure 17 initially coincides with the radius OB. It is then rotated about the centre, O, of the circle until it returns to its starting point. During the rotation the angle CÔB changes and during a complete rotation CO is said to have rotated through an angle of 360°. Other angles are defined as fractions of 360°. A right-angle is a quarter of a rotation and is 90°. Half a rotation, which takes OC as far as the other half of the diameter, is a rotation through 180°—thus the angle on a straight line is 180°.

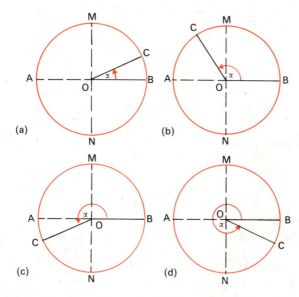

*Figure 17   The angle α measures how much the line OC has been rotated from OB: (a) for between 0° and 90°; (b) for between 90° and 180°; (c) for between 180° and 270°; (d) for between 270° and 360°.*

# 4  AREAS

The problem of estimating areas of surfaces when they are not convenient squares or rectangles is a common one in some professions. It is necessary, for example, in quantity surveying, or in estimating the areas of districts in cities and towns (remember the figures in Town Planning) from which the population density (people per square kilometre) can be calculated.

As a first step, it is as well to be able to calculate the areas of simple *geometrical* shapes because you may be able to fit such shapes together to produce a good approximation to the required area (Figure 18).

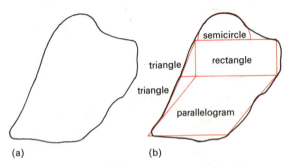

(a)                    (b)

*Figure 18   (a) An irregularly shaped site taken perhaps from a land survey. (b) The fitting together of geometrical shapes to produce a 'model' of the site whose area can now be easily calculated.*

I assume you already know that the area of a square or a rectangle is the product of the lengths of two adjacent (i.e. neighbouring) sides, as shown in Figure 19. The area of a circle of radius $r$, as already stated, is $\pi r^2$.

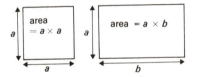

*Figure 19   The area of a square and of a rectangle.*

## 4.1  Area of a parallelogram

Figure 20(a) shows three different parallelograms drawn between the same parallel lines. The sides lying along the common parallel lines are all the same length, $w$, but clearly the lengths of the side running between the parallel lines differ from one parallelogram to another. They are all of the same area. How can we prove this?

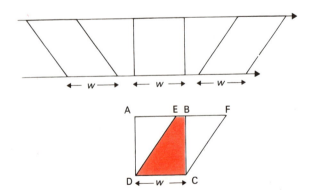

*Figure 20   (a) Three parallelograms of equal area. (b) A diagram by means of which their equality of area can be proved.*

Figure 20(b) shows a rectangle, ABCD, and one of the parallelograms drawn on the same base line DC.

A further word about names and labelling conventions. Closed geometric figures with three or more sides are called polygons. Their vertices are labelled with letters as before and a polygon is named by listing its vertices in either clockwise or anticlockwise order. *They should not be listed out of sequence.* Rectangles and parallelograms for example are four sided polygons. In Figure 20(b), ABCD is a rectangle and EFCD is a parallelogram.

To return to the question of the area of a parallelogram. What I want to do is to prove that the area of ABCD is the same as the area of EFCD.

ABCD can be thought of as the sum of two areas $AED + EBCD$; similarly, EFCD can be thought of as the sum of two areas $BFC + EBCD$. The area of EBCD is common to both the rectangle and the parallelogram, so if I can prove that triangles AED and BFC are congruent, I will have achieved what I set out to do—congruent triangles obviously have equal areas because they can be superimposed.

Looking at Figure 20(b)

$AD = BC$    (opposite sides of a rectangle)

$ED = FC$    (opposite sides of a parallelogram)

Because $AB = CD$ (opposite sides of a rectangle) and $CD = EF$ (opposite sides of a parallelogram) $AB$ must equal $EF$. If you subtract $EB$ from $AB$, you are left with $AE$, and if you subtract $EB$ from $EF$ you are left with $BF$. Therefore

$AE = BF$

All three sides of one triangle are therefore equal to their counterparts in the other; so the triangles are congruent and equal in area. Add each triangle in turn to EBCD and you have shown that the areas of the rectangle and the parallelogram are the same, which is what was required.

The area of the rectangle, however, is the length of its base, CD, multiplied by the perpendicular distance between the parallel lines forming it, namely AD—called *its perpendicular height.* Thus the area of the rectangle is $CD \times AD$. Equally, the *area of a parallelogram is the length of its base times its perpendicular height.*

All three parallelograms in Figure 20(a) have the same base, $w$, and the same perpendicular height, that is, the same distance between the parallel lines. Hence, they all have the same area.

## 4.2  Area of a triangle

Consider any triangle, for example, the one drawn in Figure 21(a): what is its area?

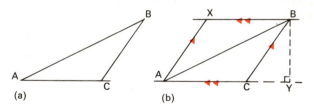

(a)    (b)

*Figure 21   The area of triangle ABC is half that of the parallelogram ACBX.*

You can see how to answer this by constructing a parallelogram around ABC, as shown in Figure 21(b). AX is drawn parallel to BC, and BX is drawn parallel to AC.

The area of the parallelogram ACBX is (length of base) × (perpendicular height), so if you could prove that ABC is half the area of this parallelogram, you would be able to state the following important result:

area of a triangle $= \frac{1}{2}$ (base) × (perpendicular height)

**SAQ 4**

Try to establish that triangles ABC and ABX are congruent and hence prove that the area of a triangle is $\frac{1}{2}$ (base) × (perpendicular height).

**SAQ 5**

The proof could equally have been carried out with BC regarded as the base. Draw the necessary construction and, in particular, mark in the perpendicular height for this case.

**SAQ 6**

Figure 22 shows an irregular shape. Estimate its area as best you can.

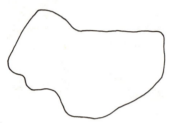

*Figure 22   An irregular shape, such as that of a city, whose area needs to be estimated.*

### 4.3   Area of the surface of cylinder

The curved surface of a cylinder can be opened out to form a rectangle as shown in Figure 23; so its area is the height of the cylinder times its circumference: that is, $2\pi r \times l$.

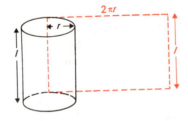

*Figure 23   The surface of a cylinder unrolled like a carpet.*

**SAQ 7**

What is the total surface area of a cylinder which is 1 m long and 40 cm in diameter, if it is closed at both ends? If it is made of metal sheet (whose thickness you may neglect) weighing 10 kilograms per square metre, how much does the cylinder weigh?

# 5 FURTHER GEOMETRICAL IDEAS AND HOW THEY CAN BE USED

## 5.1 The meaning of converse as applied to parallel lines

Consider the two straight lines AB, CD shown in Figure 24. If the lines are parallel then $X\hat{O}B = O\hat{P}D$ (corresponding angles).

You can turn this statement around and say that if $X\hat{O}B = O\hat{P}D$ it follows that AB is parallel to CD. This second statement is called the *converse* of the first. In this particular case, the converse is true, although some converses are false: for example, a zebra is something striped, but 'something striped' is not necessarily a zebra: it might be a wasp.

Let me consider the result you proved in SAQ 3; that opposite sides of a parallelogram are equal in length. The converse of this is: if the opposite sides of a four-sided polygon are equal in length then it is a parallelogram. Actually, this converse is also true, but it cannot be assumed from SAQ 3; it needs an independent proof. Later I shall come to other examples of converse theorems, which will be repeated in the summary.

**converse**

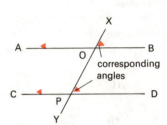

*Figure 24  If the corresponding angles $X\hat{O}B$ and $O\hat{P}D$ are equal, the lines AB and CD are parallel.*

### SAQ 8

(a) Prove the converse of SAQ 3 is true by creating a pair of congruent triangles.

(b) What is the converse of the proposition concerning alternate angles in Section 3.1?

**SAQ 8**

## 5.2 Ratios, similar triangles and maps

If two roads make an angle of 30° with each other, we expect the same angle to occur on an ordnance survey map of the area. In fact, this type of map is a model in which all angles are correctly reproduced. If A', B' and C' are three towns, they are represented on a map as A, B and C such that the corresponding angles are equal. I have tried to illustrate this in Figure 25. However, in order to avoid using a sheet of paper several miles across, it is necessary to use different scales to represent the triangle of towns A', B' and C' and the triangle ABC on the map. Although I need two symbols, A and A', to distinguish the angle on the map from the one of the landscape, I should like to use a single symbol for both of them to illustrate that they are equal angles. Since I am running out of English letters, I shall use a new notation in Figure 25 for the size of the two angles, namely α (the Greek letter alpha). β (beta) and γ (gamma) are used in a similar way.

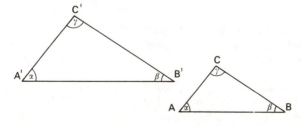

*Figure 25  Three towns A', B', C' are represented by three points A, B, C on an ordnance survey map. The angles are preserved during the mapping, but the separations (e.g. A'B') of the towns are reduced, on an actual map, by 50 000.*

23

One of the sets of information which is *not* sufficient to establish that two triangles are congruent is the specification of two (or three) angles.

If you know that two angles are the same in two triangles, can you be sure the third is too?

Yes, because in both cases, the three angles add up to 180°.

If two triangles have all their angles the same, they are called *similar triangles*. You will see that they have a second important property, namely that corresponding sides are scaled by the same factor. Before trying to prove this property, let us consider what it means in terms of our map.

**similar triangles**

Many of the latest ordnance survey maps have a scale of $1:50\,000$ ($2\,cm$ to the kilometre or about $1\frac{1}{4}$ inches to the mile). This means that distances on the landscape are reduced by a factor of $50\,000$ before drawing the map. The essential point is that only one scale is used for all distances. Thus the three towns $A'$, $B'$, $C'$ are mapped on to A, B, C in such a way that

$$AC = \frac{A'C'}{50\,000}$$

$$AB = \frac{A'B'}{50\,000}$$

$$CB = \frac{C'B'}{50\,000}$$

The first equation can be rewritten by dividing both sides by $A'C'$, thus

$$\frac{AC}{A'C'} = \frac{1}{50\,000}$$

The fraction on the right-hand side is often called the *ratio* of the numbers 1 and $50\,000$. Similarly, the left-hand side is the ratio of the lengths of AC and $A'C'$. (Note that here $A'C'$ represents the length of the line, it must never be confused with $A'$ multiplied by $C'$.) Repeating the process with the next two equations gives

**ratio**

$$\frac{AC}{A'C'} = \frac{AB}{A'B'} = \frac{CB}{C'B'}$$

As far as maps are concerned, this last result is true whatever the scale: as far as geometry is concerned, in any two similar triangles, the ratios of corresponding sides are equal.

**SAQ 9**

**SAQ 9**

Draw a triangle whose sides are eight, eleven and thirteen centimetres long. Scale up one side by a factor 1.2. On this new side, construct a similar triangle, that is, one which has the same angles as the first. Convince yourself that the other two sides have also been scaled by the same factor of 1.2.

In SAQ 9 you verified by drawing that all three sides of a triangle are scaled by the same amount when constructing a second triangle similar to the first. You can also do this by means of a more general proof.

In Figure 26, I have used Greek letters to label angles and small italicized letters to denote the lengths of sides. You can see at a glance that the big triangle in Figure 26(a) is similar to the smaller one because it has the same angles $\alpha$, $\beta$, $\gamma$. If I know that one of its sides AC is twice as long as the corresponding side FD of the smaller triangle, the problem is to demonstrate that the sides AB and CB are also doubled.

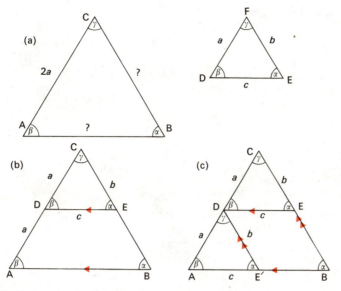

*Figure 26* To show that the sides of similar triangles are all changed by the same scaling factor. (a) ABC, DEF are similar triangles as you can see from the angles α, β and γ. AC = 2DF. The problem is to show BC/FE = AB/DE = 2 in this case. (b) Triangle DEF is fitted into triangle ABC at C: (c) Triangle DEF is now fitted into triangle ABC at A.

As a first step, lift up the small triangle and superimpose it on the other one as shown in Figure 26(b). I should like to show that $CB = 2b$ and that $AB = 2c$. You can see that DE is parallel to AB because the corresponding angles β (or α) are equal. Now repeat the superposition, but this time place the triangle at A (Figure 26(c)). On this occasion the lines DE′ and CB turn out to be parallel, because the corresponding angles γ (or α) are equal.

In Figure 26(c) the new line DE′ has created a four sided figure DEBE′.

What do we know about it?

Since we have now proved that both its pairs of opposite sides are parallel, it is a parallelogram. The properties of a parallelogram were given earlier, and you can now say that

$$E'B = DE = c$$
$$EB = DE' = FE = b$$

so that

$$\frac{CB}{FE} = \frac{AB}{DE} = \frac{AC}{DF} = 2$$

If we had started with similar triangles for which one pair of sides (AC and FD) were in the ratio 3:1, the argument would have ended with the same equation, but with 3 on the right hand side instead of 2. In principle, the proof can be extended indefinitely to triangles whose side are in any ratio (not necessarily a whole number). So, in general, if in Figure 25 triangles ABC and A′B′C′ are similar (i.e. they have the same angles) then

$$\frac{AB}{A'B'} = \frac{AC}{A'C'} = \frac{BC}{B'C'}$$

whatever the ratio $AB/A'B'$ happens to be. You can express the first equation another way by multiplying both sides by $A'B'$ and then dividing them by $AC$, the result is

$$\frac{AB}{AC} = \frac{A'B'}{A'C'}$$

Similarly, from the second pair of ratios

$$\frac{AC}{BC} = \frac{A'C'}{B'C'}$$

Finally, from the equality between the first and last ratios

$$\frac{AB}{BC} = \frac{A'B'}{B'C'}$$

The converse is also true: if the ratios of corresponding sides are all the same, the two triangles are similar. You do not need to know the proof of this result.

### SAQ 10

Drivers are expected to be able to read number plates on cars easily at a distance of about 25 metres. The letters on number plates are 8 cm high.

Now you wish to put up a road sign which drivers travelling at 70 miles per hour (say, $30 \, \text{m s}^{-1}$) will be able to read with ease. If it takes three seconds to read what is on the sign and the driver must have completed reading it 100 m before he reaches the sign (so that he can respond to what he reads), how large should the lettering on the road sign be?

## 5.3   Similar triangles and the shadow pole

You can now see why it is true that if the shadow of a 2 m stick is 2 m long then the height of a tree with a 20 m shadow is also 20 m.

Since the two triangles of Figures 2(a) and 2(b) are similar

$$\frac{AB}{PQ} = \frac{AC}{PR}$$

Alternatively, multiplying both sides by $PQ$ and substituting 20 m for $AC$

$$AB = \frac{PQ}{PR} \times 20 \, \text{m}$$

or

$$AB = \frac{2}{2} \times 20 \, \text{m}$$

$$= 20 \, \text{m}$$

### SAQ 11

Draw a diagram representing the situation in which the shadow of the 2 m stick is 1 m long and the shadow of the tree is 3 m long. How tall is the tree?

### SAQ 12

The same vertical tree is found to be on ground that slopes upwards away from the Sun at an inclination of 10° to the horizontal. The shadow of the stick (held vertically) at the time of measurement is 1.5 m long and the shadow of the tree is 4.5 m long. Draw a diagram representing the situation and show that the properties of similar triangles still apply.

## 5.4   The theorem of Pythagoras

Pythagoras' theorem is probably the most famous of all geometrical theorems. It can be shown to be true by using the properties of similar triangles. This theorem states that for any *right-angled triangle* (i.e. a triangle with one angle equal to 90°) *the square of the length of the hypotenuse* (i.e. the side opposite the right-angle) *equals the sum of the squares of the lengths of the other two sides.* That is, in Figure 27(a)

$$(BC)^2 = (AB)^2 + (AC)^2$$

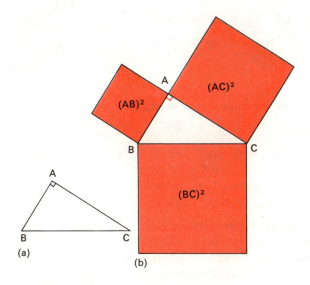

*Figure 27   Pythagoras' theorem:* $(AB)^2 + (AC)^2 = (BC)^2$.

Expressed in another way, the areas of squares drawn on the two shorter sides of the triangle add together to equal the area of the square drawn on the hypotenuse.

**SAQ 13**

If ABC in Figure 28 is a right-angled triangle and a line is drawn from A at right-angles to the opposite side, meeting it at D, show that ACD, BAD and BCA are similar triangles. (Notice the useful device of choosing the order of the vertices in such a way that the corresponding sides are obvious, i.e. AC, BA and BC are hypotenuses in each case.)

SAQ 13

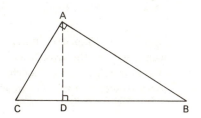

*Figure 28   See SAQ 13.*

SAQ 13 is an important step in the proof of Pythagoras' theorem. Since BCA and BAD are similar triangles (Figure 28)

$$\frac{BC}{BA} = \frac{BA}{BD}$$

(you can check this without a diagram now that the vertices are ordered correctly). Multiplying both sides by $BA$ and then by $BD$ gives

$$(BA)^2 = BC \times BD \qquad\qquad (1)$$

Similarly, since BCA and ACD are similar,

$$\frac{BC}{AC} = \frac{AC}{DC} \qquad\qquad (2)$$

and

$$(AC)^2 = BC \times DC$$

Adding equations (1) and (2) together gives

$$(BA)^2 + (AC)^2 = BC \times BD + BC \times DC$$
$$= BC(BD + DC)$$

But BD + DC is the line BC. Therefore

$$(BA)^2 + (AC)^2 = (BC)^2$$

This proves Pythagoras' theorem. It can also be proved by using appropriate constructions and comparing areas using some of the results we found earlier, but one proof is enough.

If, in Figure 28, $AB = 4\,\text{cm}$, $AC = 3\,\text{cm}$, how long is BC?

$$3^2 + 4^2 = 9 + 16$$
$$= 25$$

The converse of this theorem is: if $(BC)^2 = (BA)^2 + (AC)^2$ then the angle $C\hat{A}B$ is a right-angle and it is true.

So $BC = \sqrt{25}\,\text{cm}$
$$= 5\,\text{cm}$$

### 5.5 Constructing right-angles and bisecting angles: properties of the isosceles triangle

The presence of right-angles in a modern building shows the need to be able to construct right-angles accurately, otherwise, prefabricated parts like doors and windows would not fit properly. A small protractor is nowhere near accurate enough for setting out the foundations of such a building.

You may have noticed that in setting SAQ 13, I remarked that the line AD was to be drawn perpendicular to BC, though I did not explain how to do this. In this section, I want to show how right-angles can be drawn and how you can *bisect* angles, (i.e. divided them into two equal parts) partly because one way of constructing a right-angle is to bisect the angle of 180°.

The oldest method of constructing a right-angle makes use of the converse of Pythagoras' theorem.

### SAQ 14

A triangle with sides of length 3, 4 and 5 units (any unit will do, centimetres or inches or feet, etc.) has a right-angle between the sides of 3 and 4 units. Use this fact and a compass to construct a right-angle at the end, A, of the line AB in Figure 29.

SAQ 14

A ———————————————— B

*Figure 29  Constructing a right-angle at A. See SAQ 14.*

The method of constructing right-angles given in SAQ 14 is believed to have been used in the construction of pyramids, long before the time of Pythagoras. Before considering the other methods it is necessary to learn how to bisect an angle. Given an angle $\hat{A}$, as shown in Figure 30(a), which is

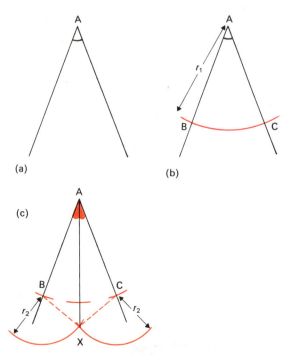

(a)     (b)

(c)

*Figure 30  Bisecting an angle. In the triangles AXB and AXC, AX is common, $AB = r_1 = AC$ and $BX = r_2 = CX$.*

to be bisected, place a compass point at A and draw part of a circle so that it intersects both lines forming the angle. Call the points of intersection B and C as in Figure 30(b). Now place the compass point on B and C in turn, and with the radius set at any value greater than a half BC, draw two more arcs of a circle which intersect at X. Then the bisector we want is the line AX (Figure 30(c)).

To prove this, consider the triangles ABX and ACX. All three sides of one triangle are equal to the corresponding sides of the other; so the triangles are congruent. Hence $B\hat{A}X = C\hat{A}X$.

**SAQ 15**

(a) Let the angle Â, which you have learned to bisect, increase until it becomes 180°. Carry out the same sequence of constructions and so construct a right-angle at the point A.

(b) Prove that the angles opposite the equal sides of an isosceles triangle are themselves equal. Hint: look at the figure you have produced in part (a).

(c) In any isosceles triangle, prove that the line drawn from the upper vertex (i.e. the one not opposite the equal sides) to the mid-point of the base (i.e. the side which is unequal to the other two) is the bisector of the angle at the vertex.

(a)

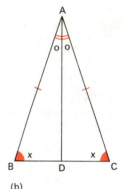

(b)

*Figure 31* (a) *An isosceles triangle,* (b) *with the vertex bisected by line AD.*

Isosceles triangles have still more properties apart from those introduced in SAQ 15. Figure 31(a) shows such a triangle. If the angle Â is bisected as shown in Figure 31(b), the two smaller triangles are congruent because AD is common, $AB = AC$ and the angles CÂD, BÂD are equal by construction. This means that the other three features of the triangles are also equal, which implies two new results. If the line between the equal angles of an isosceles triangle is considered as its base, then *the line that bisects the angle opposite the base also bisects the base and is perpendicular to it.*

You can use the second result to draw a line from a given point at right-angles to a straight line that does not pass through A. As indicated in Figure 32(a), the problem is to draw a line from a given point A at right-angles to the line PQ. First, place a compass point on A and draw the arc of a circle to cut PQ twice—at B and C, as shown in Figure 32(b). Then, as before, draw arcs of circles from both B and C so that they intersect at X, as in Figure 32(c). AX is the bisector of the angle BÂC as we proved earlier. Since the triangle ABC is isosceles, AD is perpendicular to its base BC.

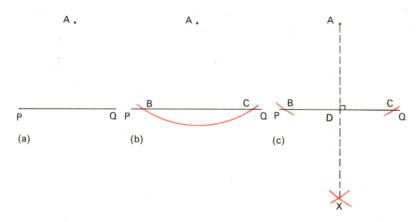

*Figure 32  Drawing a line from a point at right-angles to a given line.*

The bisection of an angle enables you to glean one further fact from shadow-pole observation: you can determine the direction in which due north lies. The length of the shadow shortens during the morning and lengthens towards the afternoon and evening. The shadow is shortest at noon and points due north (in the northern hemisphere), so if you find the direction of the shortest shadow, you can find where north lies. Around mid-day, however, the shadow does not vary much in length as its direction changes so it is not easy to find due north accurately this way. A better method is to note the directions, OX and OY, of two shadows, one in the morning and one in the afternoon, when their lengths are exactly equal (Figure 33(a)). If these two shadows are marked B and C, as indicated in Figure 33(b), with the pole labelled O, $OB = OC$ and triangle OBC is an isosceles triangle. Bisecting BÔC on the ground, perhaps using a length of string as a compass, will give you the northerly direction quite accurately.

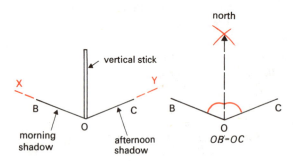

*Figure 33   Finding due north from the shadows of a vertical stick.*

## 5.6   Some properties of a circle and a sphere

### The angle in a semicircle

A triangle which lies inside a circle so that each vertex is on the circumference of the circle is said to be *circumscribed* by the circle. Such triangles in which one side is a diameter of a circle have a property which links triangles, right-angles and circles. This property is thought to have been used in the past as a means of directly checking the layout of buildings and in surveying.

Figure 34 shows a triangle drawn with the side AB as diameter of a circle and point C chosen anywhere on the circle. Because all radii of the circle are of equal length, triangles ACO and BCO are both isosceles. Thus, triangle ACO has two equal angles, $\alpha$, and triangle BCO has two equal angles, $\beta$.

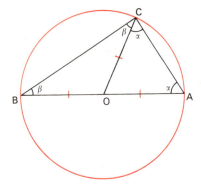

*Figure 34   Diagram for proving that the angle in a semi-circle BCA is a right-angle.*

Summing the internal angles in the triangle ACB

$$\alpha + (\alpha + \beta) + \beta = 180°$$

so

$$2\alpha + 2\beta = 180°$$

and therefore

$$\alpha + \beta = 90°$$

In other words *the lines joining any point on a circle to the ends of any diameter must be at right-angles to each other.*

The angle AĈB is often called the angle in a semi-circle. Thus an angle in a semicircle is a right-angle.

### SAQ 16

SAQ 16

Figure 35 shows a triangle OAB fitted into a circle centred on O. By the use of compasses the length AB has been set equal to the radius OA.

(a)   What name is given to triangles like AOB?

(b)   What are the values of the angles AÔB, OB̂A and BÂO?

(c) What is the value of the angle BĈO, C being the point at which the extension of the line AO cuts the circle?

(d) Find an expression for the length of BC in terms of the radius $OA = r$

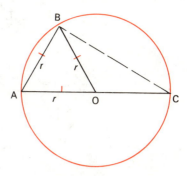

Figure 35   See SAQ 16.

## The sphere

Much of the discussion about shadow poles has been in terms of shadows being cast on the flat surface on the Earth. Actually, of course, if the ground were perfectly smooth, it would be a slightly curved surface because the Earth is more or less spherical.

I have not so far defined a *sphere*, having assumed that your everyday knowledge is sufficient. To be more formal, however, I can define it as a surface in three dimensions on which every point is a fixed distance from a given point, called the centre. A circle and a sphere are closely related. A solid sphere sliced through by a plane will always expose a flat circular surface. A good example is the Earth, which is approximately spherical (Figure 36). In Figure 37. you can see the plane section through the centre C,

**sphere**

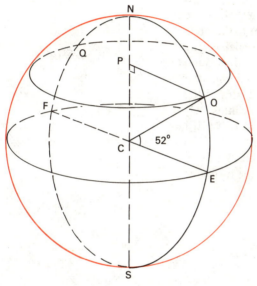

Figure 36   *A three dimensional spherical model of the Earth: the point O represents Oxford, C represents the centre and N and S represent the geographical poles.*

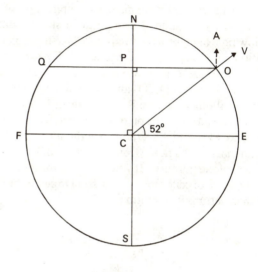

Figure 37   *A plane section through N, S and O is a circle. The angle OĈE = 52° is the latitude of Oxford. The vertical direction OV is in line with the centre C.*

the geographical poles N and S and a typical point O, such as Oxford, in the northern hemisphere. The diameter NS is called the polar axis. If a second diameter EF is drawn at right-angles to the polar axis (i.e. joining two points E and F on the equator), the angle OĈE is called the latitude of the point O. Latitude varies from 0° on the equator to 90° at the poles. At Oxford it is 52° north. Figure 38 shows a plane slice through the straight line OPQ and perpendicular to the axis NS. This is a disc whose boundary is called the *circle of latitude* through O. Our day of twenty-four hours is produced by the Earth spinning on its axis NS, causing the point O to move round the circle in the same time.

## The sundial

### Study comment

It is not essential for you to reproduce or understand everything in this section. You will not be examined on the three-dimensional geometry required for the sundial.

TV3, *Nothing New Under the Sun*, is devoted to a discussion of various sundials. Its purpose is to introduce you to some solid geometry as well as to

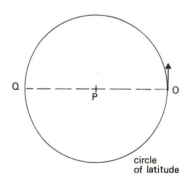

Figure 38   *A plane section through O at right-angles to NS is a disc whose boundary is the circle of latitude through O. The spinning of the Earth carries O round the circle in one day.*

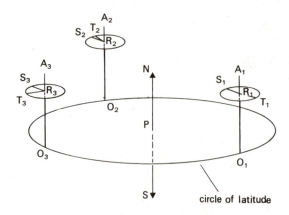

*Figure 39 How a shadow moves round a sundial. As the point O is carried round the circle of latitude it takes with it a shadow pole ORA (shown in three places labelled 1, 2, 3) attached to the platform of the sundial. The shadow RS always points the same way while the platform turns under it. If RT is a line drawn on the platform the angle TR̂S changes uniformly during the course of the day.*

illustrate a very important historical application of geometry. A simplified explanation of a sundial can be understood from Figure 39. Here the shadow pole or style of the sundial, OA, is shown perpendicular to the circle of latitude and supporting a circular platform (not drawn to scale). As the circle of latitude rotates around its centre P, it carries Oxford through successive positions $O_1$, $O_2$ and $O_3$. This causes the platform of the sundial to rotate about its centre R as shown in Figure 40, which illustrates the circle of latitude and the platform as seen from 'above'. As the radius $PU_1$ rotates to $PU_2$ and $PU_3$, it carries the line $R_1T_1$ to $R_2T_2$ and $R_3T_3$. If I made a mark at the point T on the platform, I would see it rotate round the centre R of the platform. If you have a gramophone, you can check this by placing a reel of cellotape or insulating tape on the outside of the turntable and then setting it in motion.

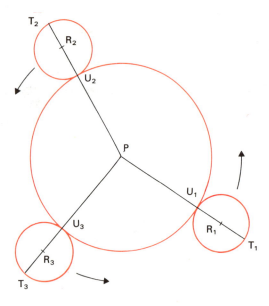

*Figure 40 This view from 'above' illustrates how the platform rotates around its centre R as it is carried round the circle of latitude. The line RT is drawn on the platform in line with the centre P. (The shadow is not shown.)*

I now want to consider what happens to the shadow. In my simplified model, the Sun is represented by a fixed source of light producing parallel rays. Since the successive positions of the style OA are parallel to each other, it follows that the shadow RS moves parallel to itself, that is, the lines $R_1S_1$, $R_2S_2$ and $R_3S_3$, which represent successive positions of the shadow RS on the platform, will always be parallel. If I were to draw the line from R to T on the platform, I should expect the shadow RS to point in a fixed direction while the line RT would rotate uniformly around it. I could therefore tell the time by measuring the angle $\hat{TRS}$. The platform could be converted into a sundial by making some regularly spaced marks on it.

In which direction is the style $O_1A_1$ pointing?

It is certainly not vertical, the vertical direction at O points upwards away from the centre C of the Earth; this is labelled COV in Figure 37. The style points along OA, which is drawn parallel to the axis NS. To work out the angle AOV between the style and the vertical note

$$A\hat{O}V = P\hat{C}O \qquad \text{(corresponding angles)}$$
$$P\hat{C}O + O\hat{C}E = 90°$$

Thus

$$P\hat{C}O = 90° - O\hat{C}E$$

Since $A\hat{O}V = P\hat{C}O$ and $O\hat{C}E$ is the latitude

$$A\hat{O}V = 90° - \text{latitude}$$
$$= 38° \qquad \text{(for Oxford)}$$

It is important that the style should point in this particular direction (i.e. parallel to the north–south axis of the Earth). If it pointed in any other direction, the successive positions of OA would no longer be parallel to each other, so that the shadow would no longer always point the same way on the rotating platform and the markings for the hours would have to vary with the seasons. The angle of the platform or dial is not so important, although it is simple to set it parallel to the circle of latitude and thus to the equator, as illustrated in Figure 39. This type is called the equatorial sundial. It has the property that the angle $\hat{TRS}$ changes uniformly, so that the markings for the hours are equally spaced.

The above discussion is based on a model according to which the Earth is supposed to spin on its axis while the Sun and the shadows stand still. When sundials were invented, however, some people supposed that they were still while the Sun went round the Earth every twenty-four hours. This older model is closer to our immediate experience as it describes the shadow moving round the sundial instead of the sundial moving under a fixed shadow.

I have simplified the above discussion by neglecting the motion of the Earth round the Sun. Although the Earth turns round once every twenty-four hours, its centre moves to a slightly new position (relative to the Sun) during the same period. Instead of remaining 'stationary' (i.e. always pointing the same way) the shadow moves a little bit each day, equivalent to an average of four minutes per day. You may think that this error is too small to worry about, many watches or clocks lose this amount or more. However, if left uncorrected for a few months, a mere four minutes a day would build up into several hours, so a correction has to be applied to the final reading.

Provided this correction (which can be looked up in tables) is applied, a good sundial is accurate to within a minute all the year round. Sundials of this type were used by some of the French railways up until this century.

### The curvature of the Earth

The curvature of the Earth is not easy to observe except at sea and the large tank shown in Unit 1 for testing model ships has sides which curve with the Earth in order that the surface of the water is a constant distance below the side of the tank. The towers of, for example, the Severn Bridge are not quite parallel—they point vertically downwards towards the centre of the Earth.

How can we measure the Earth's curvature?

When a line is drawn from the centre of a circle to the mid-point of a chord (Figure 41), it lies perpendicular to the chord. This is because the triangle OAB is isosceles, due to OA and OB being equal radii. The property that the line from the centre of a circle to the mid-point of a chord is perpendicular to the chord is independent of the length of the chord, so nothing in the above proof prevents the length of AB from being very short. If the construction is repeated, with the distance AB getting shorter each time, we reach the situation in which the point D lies very close to the point T on the circle and A, T and B are very nearly in a straight line. Eventually, a line can be drawn that is perpendicular to OD and which just touches the circle at point T. In this case, OD becomes a radius of the circle. A line like this, that just touches a circle, is called a *tangent to the circle* and it is perpendicular to the radius.

On a calm day by the sea, when you look at the horizon, you are looking along a tangent to the Earth's surface, assuming no curvature of the light-rays. A beam of light which reaches your eye just skims the surface of the sea, and so forms a tangent to it. This provides a means of estimating the distance of the horizon visible from the masthead of a ship or, if you know that distance, of estimating the radius of the Earth.

Consider a ship whose mast height I shall assume to be 30 m (Figure 42). At what distance from the ship's masthead will the horizon be visible on a clear day? Suppose the ship can be represented by the point C and its masthead by point B. Its mast height is represented by the distance $h = CB$ (not drawn to scale). The point A is where the horizon is just visible from the masthead and O represents the centre of the Earth.

I must find the distance from A to B, which I shall call $d$. If I represent the section of the Earth shown in Figure 42 as a circle, $OA = OC = r$, where $r$ is the Earth's radius. The angle $O\hat{A}B$ is a right-angle, AB being a tangent to the Earth's surface at A. By Pythagoras' theorem

$$(OA)^2 + (AB)^2 = (OB)^2$$

which in terms of the lengths shown in Figure 42 can be written as

$$r^2 + d^2 = (r + h)^2$$

To evaluate a square such as the one on the right-hand side of this equation, start by using the definition of a square, that is

$$(r + h)^2 = (r + h) \times (r + h)$$
$$= (r + h)(r + h)$$

To multiply the contents of two brackets together you have to use the rule that the second bracket must be multiplied by every item in the first bracket.

If I do this

$$(r + h)(r + h) = r(r + h) + h(r + h)$$
$$= r^2 + rh + hr + h^2$$
$$= r^2 + 2hr + h^2$$

But $(r + h)^2 = r^2 + d^2$, so

$$r^2 + d^2 = r^2 + 2hr + h^2$$

This reduces to

$$d^2 = 2hr + h^2$$
$$= h(2r + h)$$

or

$$d = \sqrt{[h(2r + h)]}$$

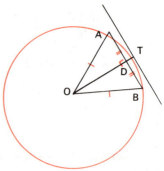

*Figure 41 A tangent to a circle is perpendicular to the radius.*

tangent to a circle

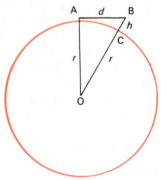

*Figure 42 A diagram representing the distance to the horizon from a ship's masthead.*

Now the Earth's radius, $r$, is much greater than the mast height, $h$, so it is nearly true that I can ignore $h$ as compared with $2r$. That is, $h(2r + h) \approx 2rh$ and so

$$d \approx \sqrt{(2rh)} \qquad\qquad (3)$$

The sign $\approx$ means 'is approximately equal to'.

For the example given, if $r = 6370 \, \text{km} = 6.37 \times 10^6 \, \text{m}$, the distance to the horizon is

$$d \approx \sqrt{(2 \times 6.37 \times 10^6 \times 30)} \, \text{m} = 19.5 \, \text{km}$$

to three significant figures.

Alternatively, if you had measured $d$ you could have estimated the radius of the Earth.

### SAQ 17

SAQ 17

Some communication systems operate with frequencies for which radio waves behave like those of light. Signals are sent across the country using towers at which signals are received and transmitted to the next tower in the chain. The Post Office Tower in London is one such tower.

Estimate how high the towers must be if their 'line of sight' distance is 50 km. Suppose, for the sake of this calculation, the stretch of country between the towers is the surface of a smooth sphere of radius 6370 km.

# 6 SURVEYING AND TRIGONOMETRY

Sometimes you may find that you do not want to use geometry because the method is not accurate enough or it may be too tedious for extensive use. Whatever the reason, you may search for alternatives. For the remainder of this unit, and in much of Unit 5, I want to develop a series of ideas which both augment and provide alternatives for the more formal geometry you have just studied.

## 6.1 Surveying using geometry

As an example of why one should look for something more than straight-forward geometry to describe the shape and position of things, I want to introduce the problem of surveying. The business of a surveyor is to make measurements of distance and direction that will relate the principal features of a landscape in sufficient detail to allow him to construct a map. The end product of his work is thus a piece of paper which shows roads, hills, trees, houses, rivers and so on. All are so placed that the horizontal distances between objects represented on the map are proportional to the distances between the real objects. Also the angles on the map are equal to the angles between the line-of-sight directions from one object to any two others.

What I want to concentrate on now is the intermediate stage between the actuality of a landscape and its representation on paper. Between the paper model of the landscape and its actuality lies another model. This is the record of measurements out of which the map will be constructed. It is the choice of measurements and the mathematics of their manipulation that will underlie most of the work of this section.

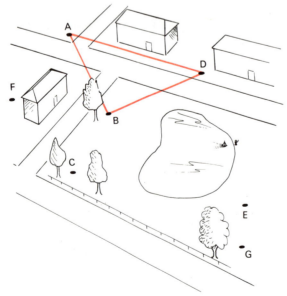

*Figure 43   How to produce a map. You do not attempt to measure everything: instead you mark some special points A, B, C, D and locate these.*

Suppose you wanted to make a map of the region shown in Figure 43. To site the points A, B, C, D, E, F and G you would equip yourself with the two basic tools of surveying—a steel chain and a theodolite. With the chain you

*Figure 44    A surveyor's theodolite. (Vickers Instruments Ltd.)*

would be able to measure the distance between the chosen points. With the theodolite (Figure 44), you would be able to measure the angles such as AD̂B and AB̂D.

In making a map, you have quite a lot of choice open to you. In the first place, you must select what you want to show, for even the most detailed of maps depict only a part of what you would see if you were to visit the area and look about you. You also have choice in the measurements you will make, for you can fix the position of the features you will map in more than one way.

To lay out positions A, B and D on your map, you might begin by measuring the distance AD. If this were 100 metres, you might represent it by a distance of 10 cm on your map. The scale of your map would then be 1:1000, or 1 in 1000. Subject to the requirement that you would want your other chosen points to appear on the map, you could draw out the line AD anywhere you chose on your paper. The position of the other points would then be fixed.

To understand this, consider the point B. You might fix it by measuring the lengths AB and DB, or by measuring the angles DÂB and AD̂B. If the lengths turned out to be 90 m for AB and 70 m for DB, you would draw an arc of 9 cm radius centred on A and another arc of 7 cm radius centred on D. The point of intersection would then correspond to the position of B on the map. Alternatively, if DÂB and AD̂B happened to be 40° and 70°, you would be able to place B using a protractor. Working from the points B and D, you could then fix the point C by measuring BC and DC or by measuring CB̂D and CD̂B. You could then go on to other points.

The two constructions I have suggested are simply applications of two of the geometrical ideas that you have met already. You can specify a triangle by fixing the lengths of all three sides or by fixing the length of one side and the values of the angles at each end of that side. Either will do to fix the position of B given the positions of A and D.

Although I have treated the two types of construction as equivalent for the drawing of a simple map, they are not of equal value to a working surveyor. You can see why this might be so by thinking of how you might make a good measurement of the distances BE and DE which pass over a small pond. Except on ideally flat ground, it is a lot easier and more accurate to measure angles than distances.

To provide the basic information for his map, the surveyor's normal approach is to make one careful measurement of length, perhaps the line AD of my figure. He then measures only angles. You can see from Figure 45 that if he takes his theodolite to the points A and D, he can make measurements of the angles denoted by $x_1$ and $x_2$ and so fix the point B. From B and D, he can then measure the angles $y_1$ and $y_2$ to fix E. From B and E, he measures $z_1$ and $z_2$ to fix C, and so on.

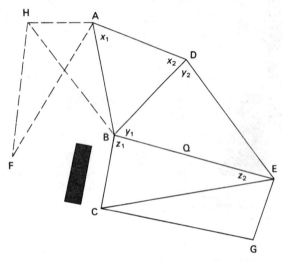

*Figure 45  This illustrates the points in the landscape (Figure 43) that have been selected for location. One length AD is measured and all of the angles.*

Only the point F of the figure evades positioning by this method. It can be seen from only A among the given fixed points. To put it on the map, you need an additional point, say H. So long as H can be seen from both A and B, and F can be seen from both A and H, the point F can be located on the map.

### 6.2   Surveying using trigonometry

The construction I have been describing is perhaps adequate for drawing a simple map, but it has one serious limitation. The precision with which points are placed depends on the construction of the drawing itself. For more serious work it would be preferable to have lists showing all the distances of the selected points one from another so that a map to any scale could readily be drawn up. To do this accurately means that some form of *calculation* must replace the graphical construction. When applied to a triangle like ADB this calculation must give the lengths AB and DB in terms of the length AD and the angles BÂD and BD̂A.

Actually, it turns out that it is not all that easy to calculate the lengths of the sides of a triangle given the length of one side and its two adjacent angles, so a series of numbers giving a tabulated solution has been constructed. Now a complete tabulation giving the lengths of the sides in all triangles with two given angles would be very long. The actual tabulation used therefore deals only with right-angled triangles and gives the ratio of the lengths of the sides in right-angled triangles.

Figure 46 shows how a tabulation like this can be applied to the map problem. The line AP is constructed perpendicular to BD. Triangles APD and APB are now both right-angled triangles. In triangle APD, AD and AD̂B are both known, so by looking up the tables for the angle AD̂B you can find the ratio of the lengths $AP/AD$ and $PD/AD$ and calculate the lengths of AP and PD from the known value of AD.

You can now use the known length of AP and the known angle AB̂D to calculate AB and BP in the right-angled triangle APB. By adding the lengths of BP and PD, you then have the length of BD, so now you know all about the triangle ABD.

The construction is one that can be continued. Since the length of BD is now known and angles DB̂E and DÊB are measured, the perpendicular distance of DQ can be calculated as well as the lengths of BQ, DE and QE.

This in turn gives the length of BE and an entry into the triangle BEC using the perpendicular ER.

It is the use of the right-angled triangle as a device for calculation that is the key idea in the branch of mathematics I shall develop. The subject to which it relates is called *trigonometry*.

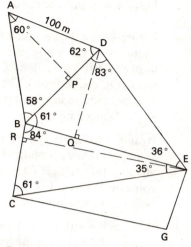

*Figure 46   The angles and the distance AD are known. One way of calculating the distances is to draw perpendiculars such as AP, DQ, ER, etc.*

### 6.3   The trigonometrical ratios for a right-angled triangle

In the constructions which I used above, one assumption is implicit. This is that in triangles of the type ABC, AB′C′, and AB″C″ of Figure 47, the sides are related by the ratios

$$\frac{BC}{AC} = \frac{B'C'}{AC'} = \frac{B''C''}{AC''}$$

and

$$\frac{AB}{AC} = \frac{AB'}{AC'} = \frac{AB''}{AC''}$$

and that these ratios are fixed for a given angle BÂC. This is because all right-angled triangles with one angle equal to BÂC are similar to each other. The converse also follows: in a right-angled triangle the ratio $BC/AC$ determines the angle BÂC.

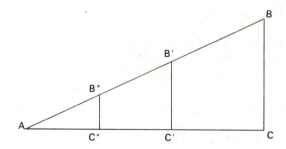

*Figure 47   Right-angled triangles sharing one angle, such as A, are all similar.*

#### SAQ 18

SAQ 18

Suppose that Figure 47 was drawn such that $BC = AC$, what would the angle BÂC be?

This shows that the ratios of sides in a right-angled triangle fix the values of the angles; put another way round, the ratios of the sides are a *property of the angle*.

### Definition of the sine ratio

Figure 48 is a device which allows me to portray all possible right-angled triangles in which one side has a length equal to the radius of the circle drawn. What I have done is to fix the line OB and then draw a circle of radius OC centred on O. The point C, wherever it is placed on the circle, can then be used to construct a triangle CQO in which the angle CÔO is a right-angle. Depending on the choice of C, Q may lie either on BO or on AO.

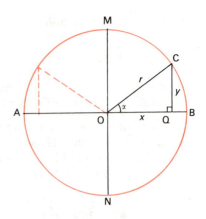

*Figure 48   A typical right-angled triangle is specified by the angle α describing the rotation of OC relative to OB.*

By moving C from near B where the value of α is near zero, I can construct triangles covering a range of values of α. When C reaches the point M, α is 90° and the triangle CQO reduces to a straight line because Q and O coincide. When C lies between M and A, the triangle opens up again. Its possible shapes, however, are a repetition of those found between B and M.

With each position of C, I can associate different values for the ratios of the lengths of the sides in the triangle COQ. In accordance with the idea that the ratios are properties of the angle, I shall give names to the ratios in terms of the angle COQ, which I have denoted α. The ratio $CQ/OC$ I shall define as the *sine* of the angle α. I shall write it as sine α when being formal, but will normally abbreviate it to sin(α) or just sin α. In both cases, the normal pronunciation is 'sign alpha'.

sine

$$\sin\alpha = \frac{CQ}{OC} = \frac{y}{r}$$

There is nothing mysterious about the values of the ratios which fix the sine of an angle. They can always be calculated from a formula that I do not want to prove here. Pocket calculators often have a facility to do this automatically. For a few particular angles, however, you can very easily work out the ratio for yourself. Figure 49(a) shows an equilateral triangle in which the three sides AB, BC, CA are equal. The line AD is constructed perpendicular to BC.

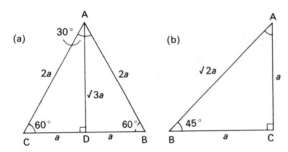

*Figure 49(a)   How to work out sines, cosines and tangents of 30° and 60° from an equilateral triangle. (b) An isosceles right-angled triangle determines the trigonometrical ratios for 45°.*

In this triangle, the angles at A, B and C are all 60° and the angle CÂD is 30°. If I take the sides of the triangle to be of length 2a, where a is a distance not specified, the lengths of CD and BD must both be equal to a. Using Pythagoras' theorem, you should be able to see that the side AD is then $\sqrt{3} \times a$, which is written simply as $\sqrt{3}a$.*

$$(AC)^2 = (CD)^2 + (AD)^2$$
$$4a^2 = a^2 + (AD)^2$$

*Note that $\sqrt{3} \times a$, $\sqrt{3}a$ or $a\sqrt{3}$ mean 'take the square root of three, and then multiply by a'. This is not the same as $\sqrt{(3a)}$ or $\sqrt{3a}$, which implies that three is first multiplied by a and then the square root of the product is taken. Compare $\sqrt{9} \times 4 = 3 \times 4 = 12$, with $\sqrt{9 \times 4} = \sqrt{36} = 6$. Where a horizontal bar is used, it should cover only the term whose square root is to be taken. You will meet these various ways of printing square roots in this course and in other mathematical texts: be sure you know the term to which the sign refers.*

Therefore

$$(AD)^2 = 3a^2$$

and so

$$AD = \sqrt{3}a$$

The sine of 60° is thus the sine of the angle $A\hat{C}D$.

$$\sin 60° = \frac{\sqrt{3}a}{2a} = \frac{\sqrt{3}}{2} = 0.866.$$

Similarly from Figure 49(b),

$$\sin 45° = \frac{a}{\sqrt{2}a} = \frac{1}{\sqrt{2}} = 0.707$$

**Study comment**

Side 1 of Disc 3 works through examples of finding trigonometric ratios for angles and illustrates how to use the trigonometric tables in your *Data Book*. Refer to your *Data Book* as you study the following sections and then listen to the disc.

**Tables of sines**

If you look at your *Data Book*, you will find on pages 6 and 7 a table headed 'natural sines'. The first column of this table is marked in degrees from 0° to 90°. In the second column, in the row corresponding to 60°, you will find the value 0.8660. This is the value of sin 60° to four significant figures.

The table is computed for angles measured to 1 minute, a minute being 1/60 of a degree. To look up the sine of the angle 25° 33′ (which means 25 degrees and 33 minutes) you look at the *row* marked 25° and then search for the *column* having the angle just less than the required one. In this case, you want the column headed 30′. The entry there is 4305 and it stands for 0.4305. To allow for the additional 3′ you now look at the *mean difference* columns where, under 3′, you will find the entry 8 in row 25. Because sines increase as the angle gets larger, mean differences are added and this 8 represents a correction to be added to the *last digit* of the number 0.4305 and thus gives for sin 25° 33′ the value 0.4313.

**SAQ 19**

Find sin 30°, sin 43° 27′ and sin 85° 6′ by using the tables of natural sines.

SAQ 19

You can also use the table in the opposite direction. Given the sine of an angle, you can find the angle. Thus the angle whose sine is 0.2588 is 15° and the angle whose sine is 0.8989 is 64° 1′. (This reverse process is often called finding the arc sine, thus arc sin 0.8989 = 64° 1′; many calculators have a key for arc sin.)

**SAQ 20**

Find the angles whose sines are 0.2893, 0.9660 and 0.7758.

SAQ 20

While the tables will always provide you with the sine of an angle, it is worth remembering that you can calculate the sine for some angles by yourself. The sines of 0°, 30°, 45°, 60° and 90° can all be calculated from triangles whose properties you know, just as was done for sin 60°.

**SAQ 21**

Find sin 30° without using tables.

SAQ 21

Road signs describing steep hills could quote the angle of the slope in degrees. Instead, the slope is likely to be quoted as 'one in five' or 'one in seven'. What the signs mean is that for every five or seven metres of movement along the road, you go up or down one metre. The sine of the angle between the line of the road and the horizontal is thus 1/5 or 1/7. For the two slopes quoted, the sines are 0.2 and 0.143, and so the angles are 11° 32′ and 8° 13′.

## The cosine ratio

cosine

Referring to Figure 48, the other ratios of importance are OQ/OC and CQ/OQ. The ratio OQ/OC is called the *cosine* of the angle α. I shall write it as cosine($\alpha$), or more usually, cos α, the pronunciations being 'co-sign alpha' or 'coz alpha'. Just as with sin α, you can calculate cos α for certain special angles; cos 45°, for example. In Figure 49(b)

$$\cos 45° = \frac{BC}{BA}$$

$$= \frac{a}{\sqrt{2}a}$$

$$= \frac{1}{\sqrt{2}} = 0.707$$

### SAQ 22

SAQ 22

Find cos 60° and cos 30° without tables.

You will find tabulated values for the cosine under the heading 'natural cosines' in your *Data Book*. They are used in the same way as the sine table with just one important difference: cosines become smaller as the angle increases, so you *subtract* the numbers in the *mean difference* columns. Thus, cos 55° 30′ = 0.5664, but cos 55° 32′ = 0.5664 − 0.0005 = 0.5659.

### SAQ 23

SAQ 23

Find cos 64°, cos 25° 24′ and cos 70° 27′. What are the angles whose cosines are 0.4226, 0.6388 and 0.8154?

## Solution of triangles containing a right-angle

If a right-angled triangle is determined uniquely by specifying two other features (not both angles) the remaining ones can be found quite easily from trigonometrical tables.

### (a) *A second angle and a side*

In Figure 46, the triangle APD was such an example; the angle $A\hat{P}D$ is a right-angle by construction, whereas I know the values of the second angle $A\hat{D}P$ and the side AD from the original measurements—AD = 100 m and $A\hat{D}P = 62°$.

To solve the triangle I have to calculate AP and PD. Now I can use tables to find

$$\frac{AP}{AD} = \sin 62° = 0.8829$$

Therefore

$$AP = 100 \times 0.8829 \, \text{m}$$

$$= 88.3 \, \text{m} \qquad \text{(to three significant figures)}$$

Similarly

$$\frac{PD}{AD} = \cos 62° = 0.4695$$

Therefore

$$PD = 100 \times 0.4695 \, \text{m}$$

$$= 47.0 \, \text{m} \qquad \text{(to three significant figures)}$$

As another example, consider a ladder standing on level ground and leaning against a vertical wall (Figure 50). When the angle $\alpha$ between the ladder and the wall is small, the ladder is a bit unstable. When the angle is large, the feet may slide away.

How much short of its full height of 4 m will the ladder reach if it is at an angle of 20° to the vertical?

The length of the ladder is 4 m, so the height of its highest point is $4 \cos 20°$ m. This represents a loss in height relative to the vertical ladder of $(4 - 4 \cos 20°)$ m or $4(1 - \cos 20°)$ m $= 4(1 - 0.9397)$ m $\approx 0.24$ m.

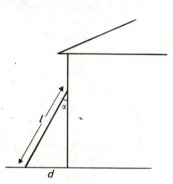

Figure 50 *The length of a ladder is given and it is at an angle $\alpha$ with the wall. Using trigonometry you can calculate the height of the ladder in this position.*

### SAQ 24

What is the furthest distance out from the wall you could place the feet of a ladder of length $l$ if you were convinced that the angle $\alpha$ should not exceed 55°? How long should $l$ be if the distance from the wall must be at least 3 m?

### SAQ 25

An aeroplane, which is losing height travels 1.5 km at an angle of 17° to the horizontal.

(a)  What is the horizontal distance travelled?

(b)  How much height does the aeroplane lose?

(c)  If the aeroplane started at a height of 4 km, how much further could it go at this angle before hitting the ground?

### SAQ 26

A motorist travelling parallel to the edge of the road at 27 metres per second (about 63 mph) has to swerve to avoid an oncoming car. In a mathematical model, it is supposed that the car turns through 3° (towards the near-side of the road) and keeps going in this direction at the original speed for a short time.

(a)  How much closer to the edge will the car be after 0.25 seconds?

(b)  If the initial distance between the car and the edge is 1.8 metres, how long can the motorist wait before pulling out of the swerve?

(b)  *Knowing two sides*

A right-angled triangle is also uniquely specified by the lengths of two of its sides. You can then calculate the angles by using the trigonometric tables in reverse. Suppose I placed a 4 m ladder against a wall and found the distance between its base and that of the wall to be 1.73 m; to calculate the angle

43

(Figure 50) between the ladder and the wall, I should proceed as follows:

$$\sin \alpha = \frac{1.73}{4}$$

$$= 0.4325$$

Therefore

$$\alpha = 25° 38' \qquad \text{(from tables)}$$

### SAQ 27

SAQ 27

A theodolite is effectively a telescope set accurately in a horizontal direction and free to rotate about a vertical axis (i.e. in the horizontal plane). (Modern theodolites can also rotate in the vertical plane, but this facility is irrelevant to this question.)

A surveyor wants to use his theodolite to measure the gradient of a steadily sloping section of road. He places his assistant on the road holding a vertical surveying pole in front of the theodolite and he reads off a height on the pole. The assistant then moves up the hill away from the surveyor a distance which he measures with his chain to be 150 m. At this point the surveyor views the pole again and finds his theodolite levelled on a point 1.3 m lower on the pole. What angle does the road make with the horizontal?

### The tangent of an angle

Consider Figure 51, which is a repeat of Figure 48. I shall use it to define a third important ratio associated with the right-angled triangle OQC, namely $CQ/OQ$ which is called the *tangent of the angle* $\alpha$. I shall write it as $\tan \alpha$, the pronunciation in this case is as you would expect—'tan alpha'.

tangent of an angle

$$\tan \alpha = CQ/OQ = y/x$$

If I write this ratio in another way, you can readily see that it is related to $\sin \alpha$ and $\cos \alpha$.

$$\tan \alpha = \frac{CQ/OC}{OQ/OC} = \frac{\sin \alpha}{\cos \alpha}$$

The 'natural tangent' tables in your *Data Book* are used like the tables of sines. There are, however, some new points to notice. Values for the tangent vary more widely than those for sine and cosine. The value of the tangent increases with the angle, having for example the value 1 at 45° (from Figure 49(b), $\tan 45° = a/a = 1$) and the value 57.29 at 89°.

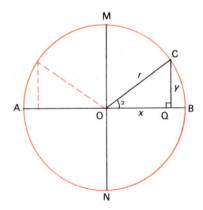

Figure 51 *The tangent of an angle is defined as* $\tan \alpha = CQ/OQ$.

In using the tangent tables, remember to be careful with the figure in front of the decimal point. For sines and cosines this number is always zero, but not for tangents. Suppose you need the tangent of 80° 6'; the shortened entry in the tables is 7297. However, if you look at the beginning of the row you will see the full entry for $\tan 80°$ is 5.6713. This indicates that you must insert a 5 as well as a decimal point to get the true result: $\tan 80° 6' = 5.7297$. If you proceed further along the row (towards the right) the entries increase steadily from 7894, 8502, 9124 to 9758, which stand for 5.7894, 5.8502, 5.9124 and 5.9758, respectively. The next entry, however, is 0405 which indicates that $\tan 80° 36'$ has now risen above 6, so its value is 6.0405.

The tables give you a warning of this by putting a bar over the first digit, writing $\bar{0}405$.

This has no mathematical significance, other publishers use a different notation. You can always work out what the missing figure should be by glancing along the row of entries as I have indicated.

Another point about the tables of tangents is that they do not contain mean differences for angles greater than 68°. One way of dealing with this difficulty is discussed later. A less accurate method is to use the nearest tabular entry.

Suppose two groups of engineers wish to bore a railway tunnel starting simultaneously from opposite sides of a small hill (Figure 52). Starting at Q, the surveyor might use his theodolite to help the engineers mark a pair of points O and C on each side of the hill in such a way as to make CQ̂O a right-angle. He could then estimate (by direct or indirect methods) the distances $x$ and $y$ (e.g. $x = 79$ m, $y = 93$ m).

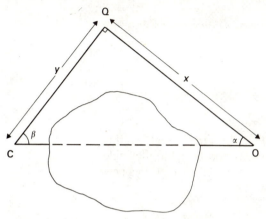

Figure 52  To ensure that the two tunnels from O and C meet under the hill, the angles $\alpha$ and $\beta$ have to be calculated. This can be done by measuring $x$ and $y$.

The next step would be to use tables of tangents to find $\alpha$ and $\beta$ as follows.

$$\tan \beta = \frac{x}{y} = \frac{79 \text{ m}}{93 \text{ m}} = 0.849$$

From tables

$$\beta = 40° 20'$$

Therefore

$$\alpha = 90° - 40° 20'$$
$$= 49° 40'$$

Armed with a knowledge of $\alpha$, the surveyor could go to O and then turn his theodolite through an angle of 49° 40' away from the direction OQ; this would provide the initial direction for the tunnel starting at O. A similar procedure could be used at C. This method is mainly of historical interest as it is similar to one suggested by a Greek engineer called Heron in the first century AD.

### SAQ 28

Find tan 5° and tan 73° 30'. What are the angles whose tangents are 0.3739 and 13.00?

(b) On the day of an equinox (e.g. 21 March or 23 September) the angle between the Sun's rays at noon and the vertical is equal to the latitude. In order to measure the latitude, an observer sets up a vertical pole of height 2.5 m and measures its shadow at noon on 21 March. If the result is 3.34 m, what is the latitude?

### Defining a radian

There is a final ratio associated with an angle that should be named along with the others. This one however is rather different in its character. Figure 53 shows a series of arcs, PS, P'S' and P″S″, linking the lines OP″ and OS″. It seems reasonable that the lengths of the arcs PS, P'S' and P″S″ will be related to the distances OS, OS', OS″, just as were the lengths of the sides of

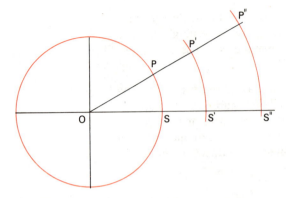

*Figure 53   To measure the angle PÔS in radians you can take any of the arcs shown and divide by its radius.*

similar triangles. Thus

$$\frac{PS}{OS} = \frac{P'S'}{OS'} = \frac{P''S''}{OS''}$$

the value of the ratio being fixed by the angle PÔS. Like the triangle's ratios this ratio is a property of the angle. This ratio is proportional to the angle measured in degrees and so is used to measure the size of the angle in a new system of units called *radians*; thus

radian

$$P\hat{O}S = \frac{\text{arc } PS}{OS}\text{radians}$$

An angle equal to one radian is shown in Figure 54; in this case the arc length is equal to the radius. In terms of degrees, this angle (1 radian) is approximately 57° 18'. I have shown it beside an equilateral (60°) triangle AOS whose sides are equal in length to the arc PS. Not surprisingly there is little difference: the construction may help you to remember roughly how large a radian is.

As with other ratios which are a property of angle, the number of radians can be evaluated for some special cases. Most important among these is the measure, in radians, of one complete rotation or 360°. In the terms of the definition, the value of this angle is the ratio of the arc of a complete circle (i.e. its circumference) to the radius of that circle. From Section 3.3, however, we know that the circumference is $2\pi r$. It follows that 360° is $2\pi$ radians and that a half a rotation (i.e. 180°) is $\pi$ radians. So 1 radian is 180° divided by $\pi$. You can check that this is 57° 18' by using your slide rule (the result will come out as a decimal number denoting the number of degrees; you can convert 57° 18' to this notation by working out 18' = (18/60)° in terms of a decimal fraction of degrees). Similarly, electronic calculators often use degrees in decimal notation rather than degrees and minutes of arc (e.g. 13° 6' would be entered as 13.1°). You can thus convert any angle from radians to degrees by multiplying by $180/\pi$ or from degrees to radians by multiplying by $\pi/180$. You do not need to remember this rule as there are conversion tables in your *Data Book*. When using the trigonometric ratios, I shall often work in terms of radians and, as a consequence, I will often quote angles in radians. In general, when the size of an angle is not denoted with a degree symbol you should assume it to be in radians. An example would be sin 45° which may be written as sin $\pi/4$.

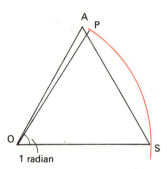

*Figure 54   Comparison of PÔS = 1 radian = 57° 18', with the angle 60° of an equilateral triangle.*

**SAQ 29**

SAQ 29

Find sin $\pi/3$ and cos $\pi/6$.

I do not want to attempt to work out the various trigonometric ratios for angles other than the few special cases you already know. The values are tabulated and you can look them up in your book of tables.

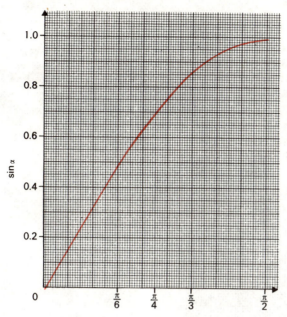

*Figure 55* *The graph of sinα again α, where the angle α is measured in radians and is restricted to the range 0 to π/2.*

It is interesting to draw graphs of sine, cosine and tangent of an angle, against the angle in radians. Figure 55 shows the graph of $\sin \alpha$ versus $\alpha$. To interpret why this curve has this kind of shape note (Figure 56) that $\sin \alpha = CQ/OC$. Since $OC$ does not vary, the height of the graph for $\sin \alpha$ is proportional to $CQ$. When C is close to B, the length $CQ$ is very small; $CQ$ increases continuously as C moves from B to M. When C is close to M, the length $CQ$ is approximately equal to $OM$. In a graph of sine against angle therefore, you should expect the value zero when the angle $\alpha = 0$, with a continuous rise to the value 1 when $\alpha = \pi/2$.

**SAQ 30**

Sketch a rough graph of cosine against angle for angles in the range 0 to $\pi/2$ (two or three intermediate points are enough).

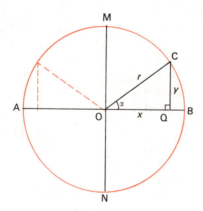

*Figure 56*

For the tangent curve, things are just a little different. The ratio this time is $CQ/OQ$ (Figure 56). As C moves away from B, two things will happen: $CQ$ becomes *longer* and $OQ$ becomes *shorter*. The ratio therefore increases as $\alpha$ increases and this happens throughout the entire arc from B to M. With C close to B, the length $CQ$ is again small so with $\alpha = 0$, as with the sine, the tangent equals zero: $\tan 0 = 0$. For larger angles, there are again the known ratios, for example $\tan 45° = \tan \pi/4 = 1$. Since the ratio must increase with angle, this means that for $\alpha > \pi/4$, the ratio for the first time exceeds one.

47

As C approaches M, the length $CQ$ is little different from $OM = OC$: the length $OQ$, however, becomes very small. Therefore, as C comes very close to M the ratio $CQ/OQ$ becomes very large indeed. It becomes so large that we cannot set up a way to count it and in this case it is denoted by a symbol, $\infty$, and given the name *infinity*.

Figure 57 shows the graph of tangent versus angle for values of $\alpha$ starting at zero and increasing towards $\pi/2$. Since we cannot set up a way of counting the value of $\tan \pi/2$, it is not represented on the graph.

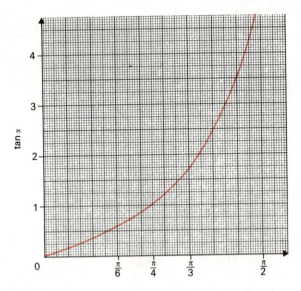

Figure 57    *The graph of the tangent of an angle (in the first quadrant). The graph breaks down near $\alpha = \pi/2$ and $\tan\alpha$ is infinite.*

### 6.4    Sines, cosines and tangents of angles outside the range 0° to 90°

You may have noticed that the entire discussion so far has been restricted to angles between 0° and 90°. The tables of the trigonometrical ratios are all given in this range. Since the angles of a triangle add up to 180°, the third angle in a right-angled triangle cannot exceed 90°, so this restriction might not appear to matter. Later on, however, I shall discuss some trigonometrical formulae which apply directly to triangles which are not necessarily right-angled triangles. Some of these triangles have obtuse angles (between 90° and 180°) for which it is useful to have definitions of sine, cosine and tangent.

The best way of thinking about larger angles is to consider the line OC of Figure 58 as it rotates in the anti-clockwise direction. So far we have restricted our attention to the quarter of the circle BOM; this is sometimes called the *first quadrant*. When $\alpha$ exceeds 90°, OC moves into the second quadrant, MOA. It is useful to have a sign convention for the different quadrants. The standard convention which is also used in graphs is to take positive directions to be above and to the right of O and negative directions to be to the left and below O. Two such lengths $x$ and $y$ are required to specify the horizontal and vertical positions of the point C relative to the axes AB and MN, respectively (Figures 56 and 58).

For the first quadrant (Figure 56), these Cartesian co-ordinates are just the same as the lengths we have been using,

$$x = OQ$$

$$y = CQ$$

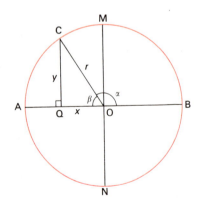

Figure 58    *For $\alpha$ greater than $\pi/2$ radians the right-angled triangle CQO falls outside the angle as shown.*

In the second quadrant (Figure 58), the $x$ co-ordinate becomes negative because C has moved to the left of O, thus

$$x = -OQ$$
$$y = CQ$$

In these equations the geometrical symbols $OQ$, $CQ$ stand for the *positive* lengths of the lines. The trigonometrical definitions now take the form of

$$\sin\alpha = \frac{y}{r}$$

$$\cos\alpha = \frac{x}{r}$$

$$\tan\alpha = \frac{y}{x}$$

where $r = OC$ and is always taken to be positive. These definitions are the same as those given in Section 6.3 for the first quadrant, but unlike them, can also be used for angles greater than 90°. In the second quadrant the new definition causes the cosines and tangents to become negative, since

$$\sin\alpha = \frac{y}{r} = \frac{\text{positive}}{\text{positive}} = \text{positive}$$

$$\cos\alpha = \frac{x}{r} = \frac{\text{negative}}{\text{positive}} = \text{negative}$$

$$\tan\alpha = \frac{y}{x} = \frac{\text{positive}}{\text{negative}} = \text{negative}$$

Notice that I have not *proved* that $\cos\alpha$ (for example) should be negative when $\alpha$ is an obtuse angle, I have only *defined* it that way by starting with the equation $\cos\alpha = x/r$. The sign conventions used in the definitions of $\sin\alpha$, $\cos\alpha$ and $\tan\alpha$ are particularly convenient for various reasons. One of these is that the graphs of sine and cosine defined this way will turn out to be smooth curves (see for example Figure 61) which are very useful in later applications.

### Trigonometrical ratios for obtuse angles

Suppose I need to find $\cos\hat{BOC} = \cos\alpha$, where $\alpha$ is the obtuse angle shown in Figure 58. I shall use the definitions given above. Thus

$$\cos\alpha = \frac{x}{r} = \frac{-OQ}{OC}$$

The right-hand side is the ratio of two of the sides of the right-angled triangle OQC. At first sight, this looks like the wrong right-angled triangle as it does not contain the angle $\alpha$ as an interior angle. You will remember, however, that it is impossible to have a right-angled triangle containing an obtuse angle. Actually, the triangle COQ is, nevertheless, very useful as you will now see;

$$\frac{OQ}{OC} = \cos\hat{QOC} = \cos\beta$$

Since $\beta$ is an acute angle (between 0 and 90°) you can look up its cosine in the standard tables and use it to find $\cos\alpha$.

Suppose I want to work out $\cos 105°$. I have

$$\alpha = 105°$$
$$\alpha + \beta = 180°$$

and so

$$\beta = 75°$$
$$\frac{OQ}{OC} = \cos 75° = 0.2588$$

so that

$$\cos 105° = \frac{-OQ}{OC} = -0.2588.$$

I now want to derive a formula for doing this a little more quickly;

$$\cos \alpha = \frac{-OQ}{OC} = -\cos \beta$$

Substituting $\beta = (180° - \alpha)$ gives

$$\cos \alpha = -\cos(180° - \alpha)$$

or if $\alpha$ is expressed in radians

$$\cos \alpha = -\cos(\pi - \alpha)$$

I can apply a similar procedure to the sines and tangents; thus

$$\sin \alpha = \frac{y}{r} = \frac{CQ}{OC} = \sin \beta$$

so

$$\sin \alpha = \sin(180° - \alpha)$$

or if $\alpha$ is in radians

$$\sin \alpha = \sin(\pi - \alpha)$$

(because $\alpha + \beta = \pi$)

Similarly,

$$\tan \alpha = \frac{y}{x} = \frac{CQ}{-OQ} = -\tan \beta$$

so

$$\tan \alpha = -\tan(180° - \alpha)$$

or if $\alpha$ is in radians

$$\tan \alpha = -\tan(\pi - \alpha)$$

You do not need to remember these formulae which are in your *Handbook*. As an example of their use, I shall find $\sin 102°$ from the tables; the formula is

$$\sin \alpha = \sin(\pi - \alpha)$$

when $\alpha$ is expressed in radians and

$$\sin \alpha = \sin(180° - \alpha)$$

when $\alpha$ is expressed in degrees. Since the tables use degrees, I have to use the second equation and let $\alpha = 102°$; thus

$$\sin 102° = \sin(180° - 102°) = \sin 78° = 0.9781$$

**SAQ 31**

SAQ 31

Find $\sin 107° 25'$, $\cos 134° 19'$, $\tan 175°$

The first two quadrants exhaust all the angles which you will need in trigonometry. This is because the largest angle of a triangle must be less than 180° (since all three add up to 180°).

**Trigonometrical ratios for angles outside the range 0° to 180°**

What happens if the line OC in Figure 58 continues to rotate beyond the end A of the diameter BOA? This is illustrated in Figures 17(c) and 17(d) which are redrawn here (Figure 59).

Although these angles are too large to occur inside a triangle it is often useful to work with their sines, cosines and tangents. It is even possible to

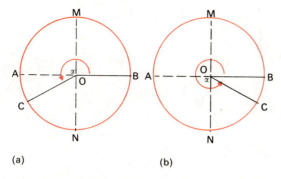

(a)                    (b)

*Figure 59(a)* $180° < \alpha < 270°$; *(b)* $270° < \alpha < 360°$.

have angles bigger than 360°. These describe rotations of more than a complete revolution. A rotation of 365° brings the line OC to the same final position as a rotation of 5°. It is also useful to have negative angles, these are used to describe rotations in the opposite direction to positive angles. The usual convention is to call anti-clockwise rotations positive and clockwise rotations negative (Figure 60).

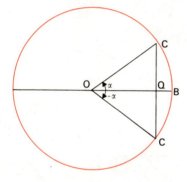

*Figure 60*  *In trigonometry, angles can be negative as well as positive; the sign of the angle is determined by the direction of rotation as shown.*

The trigonometrical definitions for sin, cos and tan can be used for any angle whatsoever, once the position of the line OC has been determined. You do not need to know how to go through the detailed arguments, but they are very similar to those described for angles between 90° and 180°.

Figure 61 shows a graph of $\sin \alpha$ versus $\alpha$ in which all the above definitions of sine have been incorporated. It is a regular undulating curve which can be extended in both the positive and negative directions of $\alpha$ as far as required.

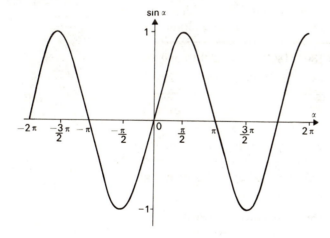

*Figure 61*  *Graph of $\sin\alpha$ over an extended range of $\alpha$. It is useful to consult this curve for checking formulae such as $\sin(\pi + \alpha) = -\sin\alpha$.*

## Repetition

The most important feature of the curve that I would like you to notice is that it repeats at intervals of $2\pi$ radians. This is due to the fact that the

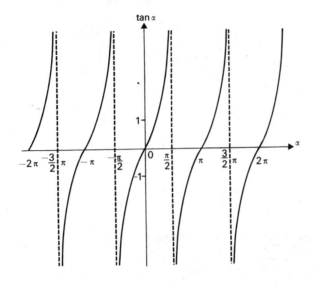

Figure 62 Graph of tan α for an extended range of α. The graph breaks down at α = −3π/2, −π/2, π/2, 3π/2,...radians when tan α becomes infinite.

position of OC (e.g. in Figure 60) is unchanged by a full revolution of $2\pi$ radians. The curve for $\tan\alpha$ (Figure 62) also repeats itself, but at half the interval, that is, every $\pi$ radians. The following formulae state the above properties concisely:

$$\sin(\alpha + 2\pi) = \sin\alpha$$

$$\tan(\alpha + \pi) = \tan\alpha$$

From Section 6.3 you know that $\tan\alpha$ goes to infinity when $\alpha$ approaches $\pi/2$ radians. As a result of the repetition just mentioned, this infinity is repeated every $\pi$ radians; this causes the graph to break down at $\alpha = -3\pi/2, -\pi/2, 3\pi/2,...$ radians as well as at $\alpha = \pi/2$ radians.

If you should ever need to work with angles greater than 90° ($\pi/2$ radians), you should consult the curves or use the following table (similar equations are given in your *Handbook*).

**Table 1   How to work out trigonometrical ratios outside the standard range 0° to 90°**

(a)   Definition of quadrants

|  | 1st quadrant | 2nd quadrant | 3rd quadrant | 4th quadrant |
|---|---|---|---|---|
| Range in degrees | 0 to 90 | 90 to 180 | 180 to 270 | 270 to 360 |
| Range in radians | 0 to $\pi/2$ | $\pi/2$ to $\pi$ | $\pi$ to $3\pi/2$ | $3\pi/2$ to $2\pi$ |

(b)   The sign of the ratio

|  | 1st quadrant | 2nd quadrant | 3rd quadrant | 4th quadrant |
|---|---|---|---|---|
| $\sin\alpha$ | positive | positive | negative | negative |
| $\cos\alpha$ | positive | negative | negative | positive |
| $\tan\alpha$ | positive | negative | positive | negative |

(c)   To use the standard tables

| 2nd quadrant | 3rd quadrant | 4th quadrant |
|---|---|---|
| $\sin\alpha = \sin(\pi - \alpha)$ | $\sin\alpha = -\sin(\alpha - \pi)$ | $\sin\alpha = -\sin(2\pi - \alpha)$ |
| $\cos\alpha = -\cos(\pi - \alpha)$ | $\cos\alpha = -\cos(\alpha - \pi)$ | $\cos\alpha = \cos(2\pi - \alpha)$ |
| $\tan\alpha = -\tan(\pi - \alpha)$ | $\tan\alpha = \tan(\alpha - \pi)$ | $\tan\alpha = -\tan(2\pi - \alpha)$ |

Standard tables give values of sin, cos and tan for angles in the first quadrant (i.e. $0 \le \alpha \le 90°$). To find sin, cos or tan for angles in the other quadrants, first substitute in the appropriate equation in part (c) of Table 1.

## 6.5 Relationships among the trigonometric ratios

### Connection between sine and cosine

Since, given the sine of an angle, you can find the angle and, given an angle, you can find its cosine, it is clear that the sine and cosine of an angle are related to one another. The relationship is simple and it is easy to work out. Look at the right-angled triangle CQO in Figure 63; you know that from Pythagoras' theorem

$$(OC)^2 = (OQ)^2 + (QC)^2$$

so, dividing both sides of the equation by $(OC)^2$

$$1 = \frac{(OQ)^2}{(OC)^2} + \frac{(QC)^2}{(OC)^2}$$
$$= \left(\frac{OQ}{OC}\right)^2 + \left(\frac{QC}{OC}\right)^2$$
$$= \cos^2\beta + \sin^2\beta$$

where $\sin^2\beta$ is the conventional way to write $(\sin\beta)^2$. The relationship $\cos^2\beta + \sin^2\beta = 1$ is a general one and is worth remembering. It is possible to prove that it holds for any value of $\beta$ including large positive or negative values (note that it does not matter what label I attach to the variable in the equation—$\sin^2\alpha + \cos^2\alpha = 1$ is also true). I shall now obtain a completely different relationship between sine and cosine which is also important.

Figure 63 shows a triangle in which $\sin\beta$ is the ratio $CQ/OC$. With a construction line CT drawn perpendicular to OM, the cosine of the angle $\alpha$ is

$$\cos\alpha = OT/OC$$
$$= QC/OC = \sin\beta$$

But $\alpha + \beta = \pi/2$, so what this relationship also shows is that

$$\cos\alpha = \sin\left(\frac{\pi}{2} - \alpha\right)$$

Now I have already shown that for any angle $y$

$$\sin(\pi - y) = \sin y$$

If I put $y = \pi/2 - \alpha$ then this becomes

$$\sin[\pi - (\pi/2 - \alpha)] = \sin(\pi/2 - \alpha)$$

The left-hand side is just $\sin(\pi/2 + \alpha)$ and the right-hand side equals $\cos\alpha$. Thus

$$\cos\alpha = \sin(\alpha + \pi/2).$$

This is a relationship that can be shown neatly in graphical terms. If you have a graph of $\sin\alpha$ against $\alpha$ as shown in Figure 64, you can construct a graph of $\cos\alpha$ against $\alpha$ from it. To do this, for each value of cosine that you

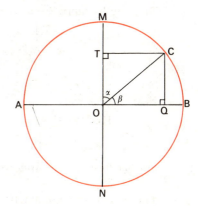

Figure 63 By relating the trigonometrical ratios for $\cos\alpha$ and $\sin\alpha$ you can show that $\cos\alpha = \sin\left(\dfrac{\pi}{2} - \alpha\right)$ and $\tan\alpha = 1/\tan\left(\dfrac{\pi}{2} - \alpha\right)$.

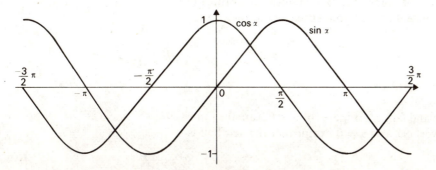

Figure 64 Comparison of the graphs of $\sin\alpha$ and $\cos\alpha$. One is just the same as the other, but shifted by $\pi/2$ radians.

want to plot, look at the value on the sine graph $+\pi/2$ radians away. You then enter this value at the angle $\alpha$ to make your cosine graph, since $\cos\alpha = \sin(\alpha + \pi/2)$. More simply, you can construct the cosine curve by sliding the sine curve back $\pi/2$ from its normal position.

In the previous section, I described how the curve for sine repeats itself every $2\pi$ radians. Since cosine is the same curve shifted sideways, it follows that it also repeats itself every $2\pi$ radians, that is, $\cos(\alpha + 2\pi) = \cos\alpha$.

### A formula involving the tangent of an angle

If you look at your tables of natural tangents, you will see that the mean differences are not quoted for angles approaching $90°$. Using Figure 63, however, you can obtain a formula which will enable you to get improved accuracy. The argument is rather similar to the previous one, you rewrite the formula for $\tan\beta$ entirely in terms of $\alpha$; thus

$$\tan\beta = \frac{QC}{OQ}$$
$$= \tan(\pi/2 - \alpha)$$

But

$$\tan\alpha = \frac{TC}{OT} = \frac{OQ}{QC}$$

Multiplying the two equations together gives

$$\tan\alpha \times \tan(\pi/2 - \alpha) = \frac{OQ}{QC} \times \frac{QC}{OQ}$$
$$= 1$$

Dividing by $\tan(\pi/2 - \alpha)$ yields

$$\tan\alpha = \frac{1}{\tan(\pi/2 - \alpha)}$$

### SAQ 32

Find $\tan 85° 45'$.

SAQ 32

### Further formulae involving sine, cosine and tangent

The various trigonometric relationships are too numerous to be remembered easily; you will, however, find them in your *Handbook*. They are things I can never remember myself, so I certainly do not want to recommend them as things that you ought to know by heart.

*Symmetry*

The simplest relationship is

$$\cos(-\alpha) = \cos\alpha$$

To interpret this concentrate on the graph of $\cos\alpha$ in Figure 64 (ignoring the second graph for $\sin\alpha$) and notice that it is symmetrical about the vertical axis where $\alpha = 0$.

The next formulae

$$\sin(-\alpha) = -\sin\alpha$$
$$\tan(-\alpha) = -\tan\alpha$$

tell us that the curves for sine and tangent (Figures 61 and 62) are the same on the right and left of the vertical axis $\alpha = 0$ except that the 'left half is minus the right half'.

*Sums of angles*

In addition there are a number of relationships by which you can calculate the trigonometric ratios for angles which are the sum of two other angles. I want to quote some of them here as you may on occasion need to look them up in your *Handbook*.

These combination laws relate the sine, cosine and tangent of the angle $(\alpha + \beta)$ to the sine, cosine and tangent of the angles $\alpha$ and $\beta$. Thus,

$$\sin(\alpha + \beta) = \sin\alpha\cos\beta + \sin\beta\cos\alpha$$
$$\cos(\alpha + \beta) = \cos\alpha\cos\beta - \sin\alpha\sin\beta$$
$$\tan(\alpha + \beta) = \frac{\tan\alpha + \tan\beta}{1 - \tan\alpha\tan\beta} \qquad \tan\alpha\tan\beta \neq 1$$

In these formulae $\alpha$ and $\beta$ can be positive or negative. They may also be equal in which case you find

$$\sin 2\alpha = 2\sin\alpha\cos\alpha$$
$$\cos 2\alpha = \cos^2\alpha - \sin^2\alpha$$

and since $\sin^2\alpha + \cos^2\alpha = 1$

$$\cos 2\alpha = 2\cos^2\alpha - 1$$
$$= 1 - 2\sin^2\alpha$$
$$\tan 2\alpha = \frac{2\tan\alpha}{1 - \tan^2\alpha} \qquad \tan^2\alpha \neq 1$$

The uses of these formulae will emerge as the course progresses.

## 6.6 Solution of triangles

I should like to return to the surveying problem I posed at the beginning of this section. Most of the basics of that problem should, I think, be clear by now. The surveyor will work from a measurement of a single length, which he calls his baseline. He then goes to his fixed points in turn and measures at least enough angles to fill out his map as a series of triangles in all of which the angles are known.

Figure 65 shows a possible minimum set of angles corresponding to a survey of the area illustrated in Figure 43. In a real survey some additional angles, say those of the triangle CDE, would be measured to give checks on the working. For present purposes however, I shall assume that the minimum set of measurements is all that need be considered.

In Section 6.3, I described how to find the unknown features of a right-angled triangle. In general terms, I am sure you know how to go about evaluating the lengths of the sides of all the triangles in Figure 65. The construction with perpendiculars of Figure 46 combined with the use of tables for sine and cosine lets you start in triangle ABD from the baseline AD of length 100 m.

In triangle ABD

$$AP = 100\sin 62°$$
$$= 88.3\,\text{m} \qquad \text{(to three significant figures)}$$
$$PD = 100\cos 62°$$
$$= 47.0\,\text{m} \qquad \text{(to three significant figures)}$$

Therefore

$$AB = AP/\sin 58°$$
$$= 104\,\text{m}$$
$$BP = AP/\tan 58°$$
$$= 55.2\,\text{m}$$
$$BD = PD + BP = 102\,\text{m}$$

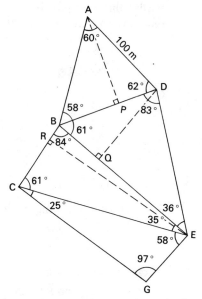

*Figure 65   This illustrates the result of a survey of the area.*

In triangle BDE, you can begin again. Here,

$$DQ = BD \sin 61°$$
$$= 89.3 \, \text{m}$$
$$BQ = BD \cos 61°$$
$$= 49.5 \, \text{m}$$

and so on.

### SAQ 33

Complete the calculation for the triangle DBE and repeat it for the triangle BEC. Do not attempt better than slide rule accuracy.

The method works, but it is rather tedious. You just begin in a triangle containing the baseline and evaluate the lengths of its sides. You then work on a new triangle for which you have evaluated one side and find the lengths of the remaining sides. You repeat this operation in the next triangle.

The calculation is clumsy because you have to evaluate the lengths of all the perpendiculars used in the construction. It would be much neater if you could calculate the lengths of the sides directly in a triangle for which one side and its adjacent angles were known.

### A triangle with one side and two angles given

I shall consider the general case in terms of triangle ABC of Figure 66. The lengths of the sides are $a$, $b$ and $c$. The line AD has been drawn perpendicular to BC, the length of AD being $p$ and the length of DB being $q$: CD is thus of length $a - q$. The triangles ABD and ACD are thus right-angled triangles.

In this diagram

$$\sin \hat{B} = p/c$$
$$\sin \hat{C} = p/b$$

By dividing, it is possible to eliminate $p$ from these equations to obtain a relationship that holds for all triangles.

$$\frac{\sin \hat{B}}{\sin \hat{C}} = \frac{b}{c}$$

or

$$\frac{\sin \hat{B}}{b} = \frac{\sin \hat{C}}{c}$$

or

$$\frac{b}{\sin \hat{B}} = \frac{c}{\sin \hat{C}}$$

Similarly, if you do this again with two other sides, you get

$$\frac{\sin \hat{A}}{a} = \frac{\sin \hat{B}}{b}$$

These two results are usually written together, and thus

$$\frac{\sin \hat{A}}{a} = \frac{\sin \hat{B}}{b} = \frac{\sin \hat{C}}{c}$$

This is often called the sine rule and is true for all triangles. It is sometimes written

$$\frac{a}{\sin \hat{A}} = \frac{b}{\sin \hat{B}} = \frac{c}{\sin \hat{C}}$$

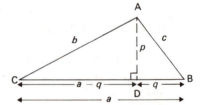

*Figure 66   This figure can be used to prove the sine rule, which is true for any triangle:*

$$\frac{\sin \hat{A}}{a} = \frac{\sin \hat{B}}{b} = \frac{\sin \hat{C}}{c}$$

*or*

$$\frac{a}{\sin \hat{A}} = \frac{b}{\sin \hat{B}} = \frac{c}{\sin \hat{C}}$$

*Note that a is opposite $\hat{A}$, b is opposite $\hat{B}$ and c is opposite $\hat{C}$.*

You must remember that $a$ is the length of the side *opposite* angle $\hat{A}$, $b$ is the length of the side opposite angle $\hat{B}$ and $c$ is the length of the side opposite angle $\hat{C}$.

In the triangle ABD of my surveying example, this formula leads readily to an evaluation of BD.

$$\frac{\sin \hat{A}}{BD} = \frac{\sin \hat{B}}{AD}$$

$$BD = AD \times \frac{\sin \hat{A}}{\sin \hat{B}} = 100 \times \frac{\sin 60°}{\sin 58°}\,\text{m}$$

$$= 100 \times \frac{0.8660}{0.8480}\,\text{m}$$

$$= 102\,\text{m} \quad \text{(to three significant figures)}$$

This confirms the earlier step by step calculation.

### SAQ 34

Figure 67 respresents a situation in which it would be difficult to estimate the height of the cliff by direct measurement. From the information given in the diagram, calculate the height, $h$, to the top of a tower on the cliff and the horizontal distance $x$.

There are two other types of problem involving triangles that can be looked at in a similar way: in one, you are told the lengths of all three sides of a triangle and want to calculate the angles; in the other, you are told the lengths of two sides and the size of the angle between them and wish to calculate the length of the remaining side.

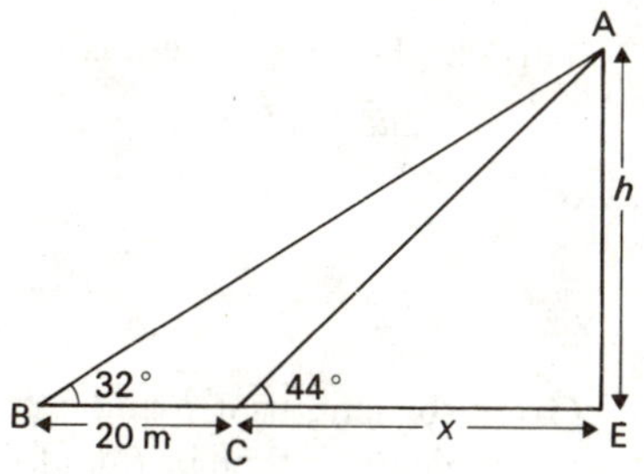

SAQ 34

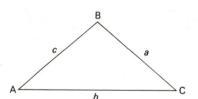

*Figure 67   See SAQ 34.*

### A triangle in which two sides and the angle between them are known

After the emphasis I have put on the measurement of angles in surveying, it might seem that this case might not arise. Such situations do happen, however, the line AC of Figure 65 providing an example.

In this case, what is the size of the obtuse angle $A\hat{B}C$?

$A\hat{B}C = 360° - (58° + 61° + 84°)$
$= 157°$

AB is already known to be 104 m and, if you have completed SAQ 33, you will know that CB is of length 113 m.

The general problem (Figure 68) is to find $b$ in terms of the opposite angle $\hat{B}$ and the two other sides $a$ and $c$. I shall just quote the formula to use, which you can also find in your *Handbook*:

$$b^2 = c^2 + a^2 - 2ac\cos \hat{B}$$

It is sometimes called the cosine (or cos) rule. (This equation look a bit like Pythagoras' theorem except that the last term is unfamiliar; this corrects for the fact that the triangle need not be a right-angled triangle. If $\hat{B} = 90°$, $\cos \hat{B} = 0$ and the formula reduces to Pythagoras' theorem.)

In terms of Figure 65, AC is given by

$$(AC)^2 = (AB)^2 + (BC)^2 - 2AB \times BC \cos(157°)\,\text{m}^2$$

From Table 1, $\cos 157°$, which is in the second quadrant, is therefore negative and equal to $-\cos(180° - 157°)$. Thus

$$(AC)^2 = (104)^2 + (113)^2 + 2 \times 104 \times 113 \cos(180° - 157°)\,\text{m}^2$$
$$= 10\,800 + 12\,800 + 23\,500 \cos 23°\,\text{m}^2$$
$$= 23\,600 + 21\,600\,\text{m}^2 = 45\,200\,\text{m}^2$$

*Figure 68   The general triangle ABC with sides of length a, b and c. If two sides, say a and c, and the angle between them, in this case $\hat{B}$, are known, the third side, b may be calculated from the formula $b^2 = c^2 + a^2 - 2ac\cos \hat{B}$. Again note that a is opposite $\hat{A}$, b is opposite $\hat{B}$ and c is opposite $\hat{C}$.*

57

Therefore, taking square roots

$$AC = 213\,\text{m}$$

(The last figure may be out by 1 or 2 since I have used a slide rule; if I had needed greater accuracy, I should have used an electronic calculator or tables).

**A triangle whose three sides are known**

The problem of finding the angles in a triangle given the lengths of its three sides would not normally come up in the course of surveying, except in making a check. A more likely possibility would be in estimating the angles of a more or less triangular piece of sheet material which was too irregular to allow the use of a small protractor.

To resolve the problem all that is needed is a rearrangement of the cosine rule. In the triangle ABC of Figure 68 you already know that

$$b^2 = c^2 + a^2 - 2ac\cos\hat{B}$$

This can be rearranged as,

$$2ac\cos\hat{B} = a^2 + c^2 - b^2$$

or

$$\cos\hat{B} = \frac{a^2 + c^2 - b^2}{2ac}$$

Given the lengths $a$, $b$ and $c$, $\cos\hat{B}$ can be calculated. By reference to tables the angle $\hat{B}$ can then be found. One advantage of this formula is that you can decide whether $\hat{B}$ is acute (i.e. between 0° and 90°) or obtuse (i.e. between 90° and 180°) according to whether the sign of the right-hand side is positive or negative; in the latter case, you should use the formula for the second quadrant (i.e. $\cos\hat{B} = -\cos(180° - \hat{B})$ in Table 1) before consulting the cosine tables.

**SAQ 35**

A triangle has sides of lengths 3 m and 4 m and the angle between these sides is 20°. Find the length of the third side.

# SUMMARY OF THE UNIT

You do not need to remember all of the detailed results obtained in Sections 1–5. What you should have learnt from them for use later in the course is:

Sections 1 5

A vocabulary, as summarized in Objective 1. (The list does not contain all the new words introduced in the unit, but it contains all the ones you need to know.) A familiarity with the way geometrical diagrams are used to model reality, how they are used and the simple shapes they deal with.

The important results obtained in Sections 3–5 include:

1   The internal angles of a triangle add up to 180°.

Section 3.1

2   Two triangles are congruent if all the sides and all the angles of the first triangle are equal to the corresponding features of the second. It is enough to show that any one of the following is true to prove that two triangles are congruent.
   (a)   All three sides are equal.
   (b)   Two sides and the angle between them are equal.
   (c)   One side and the angles at each end of it are equal.
   (d)   Any two sides are equal and the triangles contain a right-angle.

Section 3.2

3   The area of a circle is $\pi r^2$. The circumference of a circle is $2\pi r$.

Section 3.3

4   The area of a parallelogram is (the base) × (the perpendicular height).

Section 4.1

5   The area of a triangle is $\frac{1}{2}$ × (the base) × (the perpendicular height).

Section 4.2

6   Many (but *not* all) of the results in geometry have converses which can be proved to be true.

Section 5.1

7   Triangles ABC and A′B′C′ are similar if the corresponding angles are the same in each, that is, if $\hat{A} = \hat{A}'$, $\hat{B} = \hat{B}'$ and $\hat{C} = \hat{C}'$. Between similar triangles, the ratios of the corresponding sides are the same:

Section 5.2

$$\frac{AB}{A'B'} = \frac{BC}{B'C'} = \frac{CA}{C'A'}$$

The converse of this is also true: if in two triangles ABC and A′B′C′ the ratios of the corresponding sides are the same, that is, if

$$\frac{AB}{A'B'} = \frac{BC}{B'C'} = \frac{CA}{C'A'}$$

then the triangles are similar.

8   Pythagoras' theorem states that if $\hat{A}$ is a right-angle in triangle ABC, then

Section 5.4

$$(BC)^2 = (AB)^2 + (AC)^2$$

The converse is also true: if in a triangle ABC, $(BC)^2 = (AB)^2 + (AC)^2$ then $\hat{A} = 90°$.

The converses of (7) and (8) are important examples of converses that are true. There are other examples (such as the converses of (9) and (10) below) that are true, but they are less important for you to know.

9   The line that bisects the upper vertex X of an isosceles triangle XPQ, also bisects the line PQ of the triangle and is perpendicular to PQ.

Section 5.5

10   If one side AB of a triangle ABC is the diameter of the circle that circumscribes the triangle, the angle $\hat{C}$ opposite AB is a right angle.

Section 5.6

Trigonometry provides us with a method for performing accurate calculations on triangles which can be done more accurately (if necessary) and quickly than constructions with pencil and paper. The central idea is to tabulate the properties of a right-angled triangle once and for all (this has been done in the past) and to use these tables for calculations on any triangles (not necessarily right-angled).

Section 6

The main points covered are:

1    All right-angled triangles that contain the same angle (not the right-angle) are similar. Section 6.3

2    The three most important trigonometrical ratios can be defined in terms of Figure 48 and are $\sin\alpha = y/r$, $\cos\alpha = x/r$, $\tan\alpha = y/x$. These definitions apply to all four quadrants of a circle.

3    Given the angle $\alpha$, $\sin\alpha$, $\cos\alpha$ and $\tan\alpha$ can be found from the tables in your *Science Data Book*. Alternatively, they can be found from your slide rule.

4    If a triangle has been uniquely specified by three of its features (e.g. a side and two angles), then the remaining three can be calculated by using the tables just mentioned. If it is a right-angled triangle, then the tables can be used directly, but for other kinds of triangle, it depends on using the appropriate formula. Sections 6.3 and 6.6

(a)   If two angles and a side are given, use the sine rule:

$$\frac{\sin\hat{A}}{a} = \frac{\sin\hat{B}}{b} = \frac{\sin\hat{C}}{c}$$

For convenience, you may use the alternative form:

$$\frac{a}{\sin\hat{A}} = \frac{b}{\sin\hat{B}} = \frac{c}{\sin\hat{C}}$$

(b)   If two sides and the angle between them are given, use the cosine rule,

$$b^2 = a^2 + c^2 - 2ac\cos\hat{B}$$

If $\hat{B}$ is obtuse you must use the formula $\cos\hat{B} = -\cos(180° - \hat{B})$ and find the cosine of the acute angle $(180° - \hat{B})$.

(c)   If three sides are known, use the cosine rule in the form

$$\cos\hat{B} = \frac{a^2 + c^2 - b^2}{2ac}$$

If the result turns out to be negative, $\hat{B}$ is obtuse, and has then to be calculated from the formula

$$\cos\hat{B} = -\cos(180° - \hat{B})$$

The method has been illustrated by surveying. Certain points on the landscape are marked and all the angles between the various lines joining them are measured very accurately. Lengths are harder to measure accurately than angles, however only one length needs to be measured in order to determine the scale of the maps (see Figure 46). All the other distances can then be calculated by applying trigonometry to the triangles produced by joining the points. Section 6.1

The trigonometric functions sine, cosine and tangent are used outside trigonometry and in these uses the angles can have large positive or negative values. Sines and cosines have wave-like graphs which repeat every $2\pi$ radians. Section 6.4

The trigonometrical ratios for the four quadrants of a circle are all related, so the standard tables are only given for angles between $0°$ and $90°$. For other angles, the formulae in Table 1 can be used.

Whereas the curves for sine and cosine repeat every $2\pi$ radians, the tangent curve repeats every $\pi$ radians. The rate of repetition of these curves is represented in the following formulae. Section 6.5

$$\sin(\alpha + 2\pi) = \sin\alpha$$
$$\cos(\alpha + 2\pi) = \cos\alpha$$
$$\tan(\alpha + \pi) = \tan\alpha$$

$\tan\alpha$ is infinite at $\alpha = -3\pi/2, -\pi/2, \pi/2, 3\pi/2, 5\pi/2\ldots$ radians.

Sines and cosines are related by two equations

$$\sin^2 \alpha + \cos^2 \alpha = 1$$

$$\cos \alpha = \sin(\pi/2 - \alpha)$$

The trigonometric curves also have properties of symmetry about the vertical line $\alpha = 0$; see Figures 61, 62 and 64. This symmetry is expressed in the following formulae.

$$\cos(-\alpha) = \cos \alpha$$

$$\sin(-\alpha) = -\sin \alpha$$

$$\tan(-\alpha) = -\tan \alpha$$

There are formulae which relate the sine or cosine of a sum of two angles $(\alpha + \beta)$ to the sines and cosines of the individual angles, $\alpha$ and $\beta$.

$$\sin(\alpha + \beta) = \sin \alpha \cos \beta + \sin \beta \cos \alpha$$

$$\cos(\alpha + \beta) = \cos \alpha \cos \beta - \sin \alpha \sin \beta$$

A similar formula exists for tangents.

$$\tan(\alpha + \beta) = \frac{\tan \alpha + \tan \beta}{1 - \tan \alpha \tan \beta}$$

All of the results in this summary can be found in the section on trigonometry in your handbook.

# ANSWERS TO SELF-ASSESSMENT QUESTIONS

## SAQ 1

(a)  (BĜH, CĤG), (AĜH, DĤG)

(b)  (AĜE, HĜB), (AĜH, BĜE)
(CĤG, DĤF), (CĤF, DĤG)

(c)  Two pairs out of the following:
(AĜE, EĜB), (AĜE, AĜH), (EĜB, HĜB), (AĜH, HĜB)
(CĤG, DĤG), (CĤG, FĤC), (FĤC, FĤD), (FĤD, DĤG)

(d)  Two pairs out of the following:
(AĜE, CĤG), (HĜA, FĤC), (BĜE, DĤG), (HĜB, FĤD).

## SAQ 2

(a)  Draw a line 10 cm long. Use the protractor to mark out an angle of 30° at one end of the line. Draw the other arm of the angle 10 cm long, and finally join up the ends of two 10 cm lines. This is an isosceles triangle. See Figure 69(a).

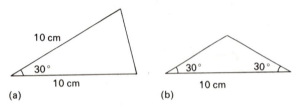

*Figure 69   Constructing triangles from given data. See the answer to SAQ 2.*

(b)  Draw a line 10 cm long. Use the protractor to mark 30° angles at each end of the line and then draw lines along these 30° angles until they meet. See Figure 69(b).

## SAQ 3

Let the parallelogram be called ABCD (as in Figure 11(a)).

So far, you have only learnt one method of proving two lines to be equal in length and that involves congruent triangles. To use this method, prepare two triangles by joining AC (Figure 70). The triangles are congruent because AC is common (that is, AC appears in both triangles), $C\hat{A}D = A\hat{C}B$ (alternate angles) and $A\hat{C}D = C\hat{A}B$ (alternate angles). Therefore $CD = AB$ and $AD = BC$.

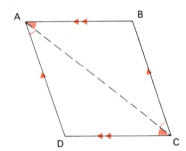

*Figure 70   Triangles ABC and ADC are congruent. This proves that $AB = DC$ and $AD = BC$.*

## SAQ 4

In the triangles ABC and ABX:
(i)    AB is common to both triangles;
(ii)   BÂC and AB̂X are equal (alternate angles);
(iii)  AB̂C and XÂB are equal (alternate angles).
Thus one side and the angles at each end of it are equal and therefore the triangles are congruent. The area of ABC is therefore half the area of the parallelogram ACBX.

The area of the parallelogram is the length of the base line AC multiplied by the perpendicular height of BY, so the area of ABC is half as much.

$$\text{area of ABC} = \tfrac{1}{2}AC \times BY$$

## SAQ 5

The area of the triangle is $\tfrac{1}{2} \times BC \times AZ$ (see Figure 71).

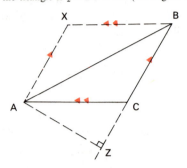

*Figure 71   See the answer to SAQ 5.*

## SAQ 6

One method of estimating the area would be to cut it out and weigh it with a sensitive balance. Another would be to use a special instrument designed for the purpose called a planimeter. However, if, as is likely, you have neither of these expensive instruments, you can use a similar method to that illustrated in Figure 18. I have illustrated one way of doing this in Figure 72, but you will be able to think of many other ones, some of which would be more accurate. Notice that

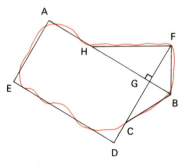

*Figure 72   See the answer to SAQ 6.*

the line CB has chopped a bit off the total area. This is compensated (approximately) by adding a bit on near the corner D. The simplified shape consists of a rectangle and three right-angled triangles giving an estimate of the area as

$$\text{area} = \text{area AGDE} + \text{area GCB} + \text{area HFB}$$
$$= AG \times GD + \tfrac{1}{2}GC \times GB + \tfrac{1}{2}HB \times GF$$

All you need do now is to measure the lengths of the lines with a ruler and substitute into the equation. Simple multiplication will then give you an estimate of the area of the original irregular shape.

The larger the geometrical shapes you use, the more difficult it is to fit them to the shape of the original area and, therefore, the less accurate is your estimate. One way of improving the accuracy would be to take a larger number of smaller shapes and find their individual areas: for example, you might use graph paper and count the number of squares, making estimates of half and quarter squares.

## SAQ 7

The area of the curved surface of the cylinder is the circumference multiplied by the height of the tank.

$$\text{area of curved surface} = 2\pi \times 0.2 \times 1 \, \text{m}^2$$
$$= 0.4\pi \, \text{m}^2$$

$$\text{area of each end of the tank} = \pi r^2$$

$$\text{Therefore area of the two ends} = 2\pi \times (0.2)^2 \, \text{m}^2$$
$$= 0.08\pi \, \text{m}^2$$

The total area is therefore $0.48\pi = 1.51\,\text{m}^2$ (to three significant figures).

The weight of the tank $= 10 \times 1.51\,\text{kg} = 15.1\,\text{kg}$

The internal surface area will be less than $1.51\,\text{m}^2$, but I have ignored this in my model because the question specifies that the thickness is negligible.

## SAQ 8

(a) The problem is to show that the opposite sides of ABCD are parallel, given that they are equal (Figure 73(a)). Join AC and compare the corresponding sides of the two triangles:

$$DA = BC \qquad \text{(given)}$$
$$DC = BA \qquad \text{(given)}$$
$$AC = AC \qquad \text{(common)}$$

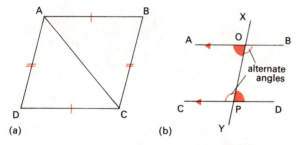

Figure 73   See the answer to SAQ 8.

Therefore, the triangles DAC and BCA are congruent (three sides equal).

It follows that they have the same angles. Let me start with $D\hat{A}C$: to find the angle equal to it in the other triangle, note that it is opposite DC, which equals BA in the other triangle which is opposite $B\hat{C}A$. Therefore, $D\hat{A}C = B\hat{C}A$ and DA is parallel to BC (alternate angles). Similarly, $D\hat{C}A = B\hat{A}C$ (also from the congruence), so that AB and DC are also parallel, which means that ABCD is a parallelogram.

(b) In Section 3.1, it was stated that if a line intersects a pair of parallel lines, then pairs of alternate angles are equal: this is illustrated in Figure 73(b).

The converse proposition is that if all alternate angles are equal, then the lines are parallel. There are two pairs of alternate angles, but it is sufficient that one pair should be equal. To see this you must remember that two angles on a straight line add up to $180°$.

Thus

$$A\hat{O}P + B\hat{O}P = 180°$$

and

$$O\hat{P}D + O\hat{P}C = 180°$$

It follows from these two equations that if $A\hat{O}P = O\hat{P}D$, then $B\hat{O}P = O\hat{P}C$. Thus, if a pair of alternate angles are equal, the second pair of alternate angles are automatically equal.

## SAQ 9

Draw out the first triangle with sides 8, 11 and 13 cm long in the way described in Section 3.2. The first side of the second triangle must be $1.2 \times 13\,\text{cm} = 15.6\,\text{cm}$ long. Draw this line. Using a protractor, at the ends of this line measure off two angles equal those at the end of the 13 cm side in the first triangle. You can now draw the other two sides of the second triangle. Measure the lengths of the two new sides of the second triangle. They should be $1.2 \times 11\,\text{cm} = 13.2\,\text{cm}$ and $1.2 \times 8\,\text{cm} = 9.6\,\text{cm}$ long.

## SAQ 10

This is a question involving similar triangles as indicated in Figure 74. If the eye is capable of reading a letter 8 cm high at 25 m, it can read a 16 cm letter at 50 m or a 32 cm letter at 100 m. So if $PQ$ is the height of a letter on the road sign and if $AP$ is the distance of the sign from the driver when he must first be able to read the sign, then, by similar triangles

$$\frac{BC}{PQ} = \frac{AB}{AP}$$

You know that $BC = 8\,\text{cm}$ if $AB = 25\,\text{m}$. You have to calculate $PQ$. What distance is $AP$? $AP$ is the distance travelled in three seconds plus 100 m. At a speed of $30\,\text{m s}^{-1}$ the distance travelled in three seconds is 90 m, so $AP = 190\,\text{m}$.

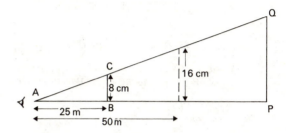

Figure 74   See the answer to SAQ 10.

Substituting these numbers in the equation gives

$$\frac{0.08\,\text{m}}{PQ} = \frac{25}{190}$$

Therefore

$$PQ = \frac{0.08 \times 190\,\text{m}}{25}$$
$$= 0.608\,\text{m}.$$

Each letter should be at least 61 cm high. Remember: $1\,\text{cm} = 0.01\,\text{m}$ and you should use the same units in your calculations.

## SAQ 11

The two triangles which represent the observed shadows, and the stick and tree which cause them, are shown in Figure 75. These are similar triangles.

$$\frac{1\,\text{m}}{2\,\text{m}} = \frac{3\,\text{m}}{\text{height of tree}}$$

Therefore, the tree is 6 m high.

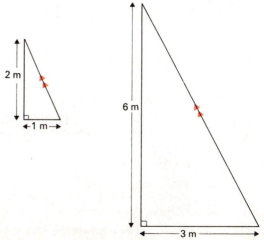

Figure 75   See the answer to SAQ 11.

## SAQ 12

The triangles representing the situation are shown in Figure 76.

ABC and DEF are similar triangles, because

(a) AB and DE are both vertical and therefore parallel, so $B\hat{A}C = E\hat{D}F$ (corresponding angles).

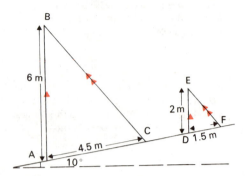

*Figure 76   See the answer to SAQ 12.*

(b) BC and EF represent beams of sunlight, so they are very nearly parallel. Therefore, $B\hat{C}A = E\hat{F}D$ (corresponding angles).

It follows that

$$\frac{AB}{ED} = \frac{AC}{DF}$$

Hence

$$\frac{AB}{2\,\text{m}} = \frac{4.5\,\text{m}}{1.5\,\text{m}}$$

Therefore $AB = 6\,\text{m}$.

## SAQ 13

The triangles concerned are reproduced in Figure 77(a). Mark the angle at C with a dot and the angle at B with a cross, as shown. To show that triangles ACD, BAD and BCA are similar, you must show that the same angles from each triangle are equal. The unknown angles are those at A, that is, $C\hat{A}D$ and $D\hat{A}B$.

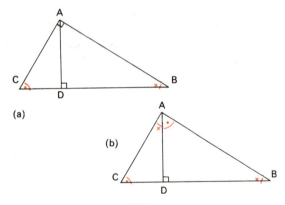

*Figure 77   See the answer to SAQ 13.*

The internal angles of triangle BCA are a right-angle, a cross and a dot and they must add up to 180°.

In triangle BAD, the internal angles must also add up to 180° and there are a right-angle and a cross, so the third angle must be a 'dot'. That is, $D\hat{A}B = D\hat{C}A$.

Similarly, $C\hat{A}D = A\hat{B}D$.

Figure 77(b) shows the angles at A appropriately marked. You can see that triangles ACD, BAD and BCA all contain the same angles and are therefore similar triangles.

Note I have named the triangles with a letter ordering that reflects the similarity of the triangles. From the order given AC, BA and BC are opposite the right-angle; AD, BD and BA are opposite the dot;

and CD, AD and CA are opposite the cross in the respective triangles. The hypotenuse is named first, the shortest side last. This makes it much easier to write down the ratios of the sides since you do not have to look back at the diagram to get it right.

## SAQ 14

To construct a right-angle using the converse of Pythagoras' theorem, you use a compass to mark out the lengths 3, 4 and 5 units. First centre your compass on A and draw an arc of a circle of radius 4 units to cut the line AB at P (see Figure 78). Then draw the arc of a circle of radius 3 units, using point A as the centre of the circle. Finally, draw another arc to intersect this one at Q, using a radius of 5 units and with the compass point placed at P. The line QA forms the required right-angle at A; that is, QA is perpendicular to AB.

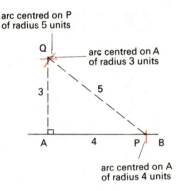

*Figure 78   Constructing a right-angle using Pythagoras' theorem and a compass (answer to SAQ 14).*

## SAQ 15

(a) Beginning with a point A on the line, first mark off equal distances from A by drawing the arcs of a circle so that they intersect the line at P and Q (see Figure 79(a)). Now, opening out the compasses to a longer radius, draw two intersecting arcs (of equal radius) centred in P and Q. If they intersect at X (Figure 79(b)) AX is at right-angles to PQ. (Triangles XPA and XQA are congruent triangles: three sides the same in each triangle.)

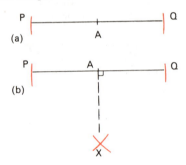

*Figure 79   See the answer to SAQ 15.*

(b) You could verify the angles are equal by constructing the triangle and measuring them. To prove that the result is always true consider a triangle XPQ with $XP = XQ$; join X to the mid-point A of PQ and note that the triangles are congruent (as in (a)) so that $\hat{P} = \hat{Q}$.

(c) $P\hat{X}A = Q\hat{X}A$ from the congruence just proved.

## SAQ 16

(a) Since all the three sides of triangle AOB are the same length, the triangle is an equilateral triangle (see Figure 6).

(b) Since AOB is also an isosceles triangle, whichever way you look at it, and since the angles at the base of an isosceles triangle are equal it follows that all the internal angles of the triangle are equal and must therefore be 60° each (their total is 180°). $O\hat{A}B = O\hat{B}A = A\hat{O}B = 60°$.

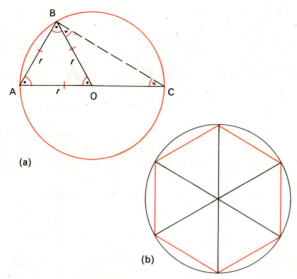

(a)

(b)

Figure 80   See the answer to SAQ 16.

(c)   In Figure 80(a), $OB = OC$ because they are radii of the circle. Therefore, triangle OBC is an isosceles triangle and, hence, $O\hat{C}B = O\hat{B}C$ as indicated by the crosses in the figure. But

$$O\hat{B}C = A\hat{B}C - A\hat{B}O$$

Since triangle ABC is circumscribed by the circle and AC is a diameter, you know that $A\hat{B}C$ is a right-angle. Also, $A\hat{B}O = 60°$. Therefore

$$O\hat{B}C = 90° - 60°$$
$$= 30°$$

As $O\hat{B}C = O\hat{C}B$, the angle $O\hat{C}B$ must also be equal to 30°.

You may have used another equally good method. $A\hat{O}B$ and $B\hat{O}C$ are angles on a straight line and, therefore, $B\hat{O}C = 180° - 60° = 120°$. Since the internal angles of a triangle add up to 180°,

$$O\hat{C}B + O\hat{B}C + 120° = 180°$$

Thus

$$O\hat{C}B + O\hat{B}C = 180° - 120$$
$$= 60°$$

Since $O\hat{C}B = O\hat{B}C$ (triangle OBC is isosceles) angles $O\hat{C}B$ and $O\hat{B}C$ must therefore be both 30°.

It is interesting to note that the construction of triangles that are congruent to OAB can be repeated to produce the shape shown in Figure 80(b). The six outer edges form a regular hexagon.

(d)   Since ABC is a right-angled triangle the theorem of Pythagoras applies to it. Thus

$$(AC)^2 = (AB)^2 + (BC)^2$$

This equation can be written as

$$(2r)^2 = r^2 + (BC)^2$$
$$4r^2 = r^2 + (BC)^2$$
$$4r^2 - r^2 = (BC)^2$$
$$(BC)^2 = 3r^2$$

Taking square roots

$$BC = \sqrt{3}r$$

## SAQ 17

Figure 81 shows a geometrical model of the situation, with the heights of the towers greatly exaggerated as compared with the radius, $r$, of the Earth. The 'tower' PQ will just be visible from the top of the 'tower' XY if QY just skims the surface of the Earth at A.

You know the relationship between XY and AY derived in equation (3): namely,

$$AY = \sqrt{(2rXY)}$$

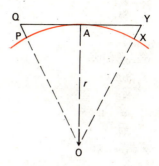

Figure 81   See the answer to SAQ 17.

You are also told the distance between the towers, QY, is 50 km, so AY is 25 km.

(Note that although the arc PA represents the distance you would measure over the ground with a tape measure, the distance QA represents what you would measure using optical surveying instruments. For the purposes of this calculation, the two distances are not significantly different over only 25 km.)

Thus

$$25 = \sqrt{(2 \times 6370 \times XY)}$$

So

$$XY = \frac{25^2}{2 \times 6370} \, \text{km}$$
$$= \frac{625}{2 \times 6370} \, \text{km}$$
$$= 0.049 \, \text{km}.$$

The towers would have to be nearly 50 m high.

## SAQ 18

If $BC = AC$, the triangle is isosceles as well as right-angled. It follows that the angles $\hat{A}$ and $\hat{B}$, which are opposite the equal sides, are themselves equal.

Since $\hat{C} = 90°$ and   $\hat{A} + \hat{B} + \hat{C} = 180°$

$$\hat{A} + \hat{B} = 90°$$

Therefore,

$$\hat{A} = \hat{B} = 45°$$

## SAQ 19

Use your table of natural sines.

(a)   Sin 30° = 0.5000 (tabular entry).

(b)   The tabular entry for sin 43° 24′ is 6871, the mean difference for 3′ is 6, so the result is 0.6877 (to check for the decimal point look at the first column).

(c)   the tabular value is 9963 so the required answer is 0.9963.

## SAQ 20

From the table of natural sines:

(a)   sin 16° 49′ is 0.2893, so the answer is 16° 49′.

(b)   sin 75° 1′ is 0.9660, so the answer is 75° 1′.

(c)   sin 50° 53′ is 0.7758; so the answer is 50° 53′.

## SAQ 21

An angle of 30° occurs in the construction shown in Figure 49(a). It follows that

$$\sin 30° = \sin D\hat{A}C$$
$$= \frac{CD}{AC}$$
$$= \frac{a}{2a} = \frac{1}{2}$$

65

## SAQ 22

Using Figure 49(a)

$$\cos 60° = \cos D\hat{B}A$$
$$= \frac{DB}{AB}$$
$$= \frac{a}{2a} = \frac{1}{2}$$

$$\cos 30° = \cos D\hat{A}B$$
$$= \frac{DA}{AB}$$
$$= \frac{\sqrt{3}a}{2a} = \frac{\sqrt{3}}{2}$$

## SAQ 23

Use your table of natural cosines.

(a)  $\cos 64° = 0.4384$ (in table).

(b)  The entry for 25° 24′ is 9033 which, from the first column of the table, is to be interpreted as 0.9033.

(c)  The entry for 70° 24′ is 3355 and the difference for 3′ is 8, which you must subtract to give an answer 0.3347.

From the table of natural cosines: $\cos 65° = 0.4226$, so the answer is 65°. $\cos 50° 18′ = 0.6388$, so the answer is 50° 18′. $\cos 35° 18′ = 0.8161$, but this is 0.0007 too high. 0.0007 corresponds to a mean difference of 4′ which must be *added to* 35° 18′ (a *larger* angle has a smaller cosine), hence the answer is 35° 22′.

## SAQ 24

The furthest distance from the wall, $d$, is related to the maximum angle 55° by

$$\frac{d}{l} = \sin 55°$$

Therefore

$$d = l \sin 55° = 0.819l$$

If $d$ has to be at least 3 m, the minimum value of $l$ is given by

$$0.819 l_{min} = 3\,\text{m}$$

Therefore $l_{min} = 3.66\,\text{m}$

## SAQ 25

(a)  The horizontal distance required is labelled $d$ in Figure 82. To calculate it, note

$$\frac{d}{1.5\,\text{km}} = \cos 17°$$

Therefore

$$d = 1.5 \times \cos 17°\,\text{km} = 1.43\,\text{km}$$

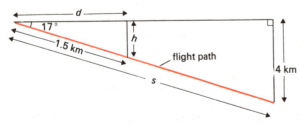

*Figure 82   See the answer to SAQ 25.*

(b)  The vertical distance lost during the same interval is $h$, where

$$\frac{h}{1.5\,\text{km}} = \sin 17°$$

Therefore

$$h = 1.5 \times \sin 17°\,\text{km}$$
$$= 0.439\,\text{km}$$

(c)  Let the *total* distance to reach the ground be $s$, then

$$\frac{4\,\text{km}}{s} = \sin 17°$$

Therefore

$$s = \frac{4}{\sin 17°}\,\text{km}$$
$$= 13.7\,\text{km}$$

The plane will already have travelled 1.5 km, so it has 12.2 km further to go.

## SAQ 26

(a)  The motorist will travel from M to B (Figure 83) in the first 0.25 seconds. Here

$$MB = 27 \times 0.25\,\text{m}$$
$$\frac{AB}{MB} = \sin 3°$$
$$= 0.0523$$

and hence

$$AB = MB \times 0.0523$$

Therefore

$$AB = 27 \times 0.25 \times 0.0523\,\text{m} = 0.353\,\text{m}.$$

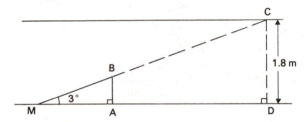

*Figure 83   See the answer to SAQ 26.*

(b)  He must pull out of the swerve before reaching C where

$$\frac{1.8\,\text{m}}{MC} = \sin 3°$$
$$= 0.0523$$

Multiplying by $MC$ and dividing by 0.0523 gives

$$MC = \frac{1.8\,\text{m}}{0.0523}$$
$$= 34.4\,\text{m}$$

Since he travels at 27 metres per second, the time he has is

$$\frac{34.4}{27} = 1.27\,\text{seconds}$$

## SAQ 27

By moving 150 m along the road the pole has been 'raised' 1.3 m. The sine of the angle with the horizontal is therefore $1.3/150 = 0.00867$.

By using the sine tables in your *Data Book* in reverse, the required angle is found to be 30′.

## SAQ 28

(a)  $\tan 5° = 0.0875$.

The entry for 73° 30′ is 3759. Keeping to the same row, look at the first column (headed 0′), where the entry is 3.2709. This indicates that the required answer is 3.3759.

tan 20° 30′ = 0.3739, so the answer is 20° 30′.

tan 85° 36′ = 13.00, so the answer is 85° 36′.

(b) Draw a diagram of the pole at right-angles to its shadow, as illustrated in Figure 84. According to the question, at an equinox the angle $\alpha$ is equal to the latitude. Now

$$\tan \alpha = \frac{3.34}{2.5}$$

$$= 1.336$$

Thus

$$\alpha = 53° 11′$$

and so the latitude is 53° 11′.

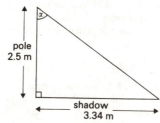

*Figure 84    See the answer to SAQ 28.*

## SAQ 29

Since

$$2\pi \text{ radians } = 360°$$

$$\pi/3 \text{ radians } = \frac{360}{2\pi} \times \frac{\pi}{3}$$

$$= 60°$$

Therefore

$$\sin \pi/3 = \sin 60°$$

$$= \frac{\sqrt{3}a}{2a}$$

$$= \frac{\sqrt{3}}{2} \quad \text{(from Figure 49(a))}$$

and thus

$$\sin \pi/3 = 0.866$$

Similarly,

$$\cos \pi/6 = \cos 30°$$

$$= \frac{\sqrt{3}a}{2a}$$

$$= \frac{\sqrt{3}}{2}$$

and therefore

$$\cos \pi/6 = 0.866$$

## SAQ 30

Start at $\alpha = 0$.

$$\cos 0 = 1$$

$$\cos \pi/6 = 0.866 \quad \text{(from SAQ 29)}$$

$$\cos \pi/4 = \cos\left(\frac{\pi}{4} \times \frac{360°}{2\pi}\right)$$

$$= \cos 45°$$

$$= \frac{1}{\sqrt{2}} \quad \text{(from Figure 49(b))}$$

Thus

$$\cos \pi/4 = 0.7071$$

$$\cos \pi/3 = \cos\left(\frac{\pi}{3} \times \frac{360°}{2\pi}\right)$$

$$= \cos 60°$$

$$= \frac{1}{2} \quad \text{(from Figure 49(a))}$$

Thus

$$\cos \pi/3 = 0.5.$$

$$\cos \pi/2 = \cos\left(\frac{\pi}{2} \times \frac{360°}{2\pi}\right)$$

$$= \cos 90°$$

When the angle $\alpha$ in Figure 56 approaches a right-angle, the length of $OQ$ tends to zero; it follows that $\cos \alpha = OQ/OP$ tends to zero, i.e.

$$\cos \pi/2 = 0$$

You now have five points on the graph of $\cos \alpha$ which you can plot roughly and join up, as shown in Figure 85.

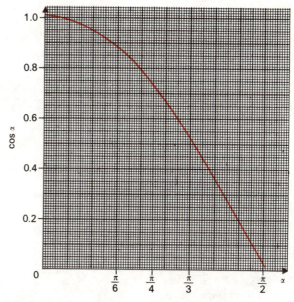

*Figure 85    Answer to SAQ 30. A sketch of the graph of $\cos \alpha$ obtained from five points.*

## SAQ 31

Sin 107° 25′ = sin(180° − 107° 25′) = sin 72° 35′. The nearest tabular value is for 72° 30′ which shows 9537, the mean difference for 5′ is 4 so the answer is 0.9541.

Cos 134° 19′ = −cos(180° − 134° 19′) = −cos 45° 41′. The table shows 6997 for 45° 36′ and 10 for the mean difference for 5′, this must be subtracted, so the answer is −0.6987.

Tan 175° = −tan(180° − 175°) = −tan 5° = −0.0875.

## SAQ 32

tan 85° 42′ = 13.30.

However, you cannot find the correction required to give you tan 85° 45′ as there are no mean differences quoted. To get a more accurate answer use

$$\tan 85° 45′ = \frac{1}{\tan(90° - 85° 45′)}$$

$$= \frac{1}{\tan 4° 15′} = \frac{1}{0.0743} = 13.46$$

## SAQ 33

Since $DQ$ is known, $DE$ should be calculated next using the right-angled triangle DQE.

$$\frac{DQ}{DE} = \sin 36°$$

67

Therefore

$$DE = \frac{DQ}{\sin 36°}$$

$$= \frac{89.3 \, \text{m}}{0.5878}$$

$$= 152 \, \text{m}$$

From triangle DQE, you can also calculate $QE$.

$$\frac{DQ}{QE} = \tan 36°$$

$$QE = \frac{DQ}{\tan 36°}$$

$$= \frac{89.3 \, \text{m}}{0.7265}$$

$$= 123 \, \text{m}$$

Since the point Q divides the side BE into two parts

$$BE = BQ + QE$$

$$= 173 \, \text{m}$$

You now have to go through a similar series of steps for the triangle BEC. From triangle BER

$$\frac{BR}{BE} = \cos 84°$$

Therefore

$$BR = BE \cos 84°$$

$$= 18.1 \, \text{m}$$

Also for triangle BRE

$$\frac{RE}{BE} = \sin 84°$$

$$RE = BE \sin 84°$$

$$= 172 \, \text{m}$$

From triangle CRE

$$\frac{RE}{EC} = \sin 61°$$

$$EC = \frac{RE}{\sin 61°}$$

$$= 197 \, \text{m}$$

and

$$\frac{RC}{EC} = \cos 61°$$

$$RC = EC \cos 61°$$

$$= 95.5 \, \text{m}$$

You could check the last result by doing it a different way.

$$\frac{RE}{RC} = \tan 61°$$

$$RC = \frac{RE}{\tan 61°}$$

$$= 95.3 \, \text{m}*$$

Since R divides the line BC

$$BC = BR + RC$$

$$= 113 \, \text{m}$$

## SAQ 34

You wish to find the height $h$ and the distance $x$. In the geometrical model of the situation represented by Figure 67, $h$ and $x$ are the lengths of the sides AE and CE of a right-angled triangle ACE and angle $A\hat{C}E = 44°$. You know that since ACE is a right-angled triangle

$$\frac{h}{AC} = \sin 44°$$

and

$$\frac{x}{AC} = \cos 44°$$

If you can use the additional information given in the question to calculate the length of AC, then it will be easy to calculate $h$ and $x$. The line AC is one side of the triangle ABC and, using the sine rule,

$$\frac{AC}{\sin 32°} = \frac{20 \, \text{m}}{\sin B\hat{A}C}$$

(Note: I have used the alternative form the sine rule, since I want to calculate a side of the triangle and this form conveniently leaves $AC$ in the numerator of the right-hand side of the equation.)

Since the angles on a straight line add up to 180°

$$B\hat{C}A = 180° - 44°$$

$$= 136°$$

and since the interior angles of a triangle add up to 180° as well

$$B\hat{A}C = 180° - (32° + 136°)$$

$$= 12°$$

Therefore

$$\frac{AC}{\sin 32°} = \frac{20 \, \text{m}}{\sin 12°}$$

and so

$$AC = \frac{20 \sin 32°}{\sin 12°} \, \text{m}$$

$$= 51 \, \text{m}$$

Thus

$$\frac{h}{51 \, \text{m}} = \sin 44°$$

$$h = 51 \times \sin 44° \, \text{m}$$

$$= 35.4 \, \text{m}$$

and

$$\frac{x}{51 \, \text{m}} = \cos 44°$$

$$x = 51 \times \cos 44° \, \text{m}$$

$$= 36.7 \, \text{m}$$

## SAQ 35

The cosine rule gives the square of the unknown length $l$

$$l^2 = (3^2 + 4^2 - 2 \times 3 \times 4 \cos 20°) \, \text{m}^2$$

$$= 2.45 \, \text{m}^2$$

Now take the square root

$$l = 1.57 \, \text{m}$$

*All numbers are quoted to three significant figures and the discrepancy in these values for RC arises from the fact that rounded results are carried forward.*

# 5. Lines, Curves and Directions

# CONTENTS

## AIMS

The aims of this unit are:

1   To introduce you to a more powerful and more general method of solving geometric problems using algebra.

2   To apply this technique to straight lines and circles.

3   To give you practice in the technique so that you will be able to set up and solve equations of straight lines and circles when they arise later in the course.

4   To show that this technique, applied to circles and straight lines, can be used to develop mathematical models relating to shopping behaviour.

5   To introduce the equations and properties of other curves that, together with the circle, are known collectively as the conic sections.

6   To extend the method of dealing with position to dealing with movement from one position to another and to introduce the topic of vectors.


## OBJECTIVES

When you have studied this unit you should be able to:

1   Distinguish between true and false statements concerning, or explain in your own words, the following terms:

| | |
|---|---|
| acceleration | origin |
| asymptote | proportionality |
| components of a vector | resolution of a vector |
| conic sections | scalar |
| co-ordinate axes | vector |
| focus | vector sum |
| intercept | velocity |

2   Explain in your own words the use of rectangular Cartesian co-ordinate axes

3   Label a point, in two or three dimensions, using symbols or numbers that relate to a given set of Cartesian co-ordinate axes (SAQ 1).

4   Calculate the length of a line between two points whose co-ordinates are specified and, hence, show whether a triangle is right-angled, given the co-ordinates of its vertices (SAQs 2 and 3).

5   Calculate the gradient of a line from the co-ordinates of two points through which the line passes, and find the equation of such a line (SAQs 3 and 9).

6   Find the equation of a line given its gradient and one point through which it passes (SAQs 7, 8 and 10).

7   State the conditions for two lines to be parallel or perpendicular; determine whether two lines are parallel or perpendicular (SAQs 4 and 10).

8   Show whether three points lie on a straight line (SAQ 5).

9   State the equation of a circle centred at the origin of rectangular Cartesian co-ordinates and the equation of a circle centred upon some arbitrary point; find the centre and radius of a circle from its equation (SAQs 11, 12, 13, 14, 16 and 17).

10   Describe a simple shopping model (SAQs 15, 16 and 17).

11  Sketch the shape of other *conic sections*: a parabola, an ellipse and a hyperbola; write down and recognize standard equations for these curves and write down the equation of a given parabola (SAQs 18, 19, 20 and 21).

12  Recognize a number of situations in which vector addition is necessary and perform such addition both graphically and by calculation, as required (SAQs 22, 23 and 24).

13  Resolve a vector of position, velocity, acceleration or force into its components and calculate the magnitude and direction of any such vector given its components (SAQs 25, 26, 29, 31 and 32).


## STUDY GUIDE

For this study week you should read Unit 5, *Lines, curves and direction*, listen to the second side of Disc 3 on *Resolving Vectors*, and complete your assignment material.

You will need to be familiar with the material in Unit 4 on the circle and the use of Pythagoras' theorem. You will also need to be familiar with the idea of representing a point of a graph by means of Cartesian co-ordinates, though it is briefly revised at the beginning of this unit. You may find it helpful to re-read Town Planning in *Modelling Themes* before you study Section 2.

You should not listen to the disc on *Resolving Vectors* until after you have read Section 4, in which this topic is introduced.

Full details of the assignment material associated with Unit 5 are given in the supplementary material.

When you have finished the unit, you can use the objectives and summary to check what you are expected to know and be able to do as a result of this week's study.

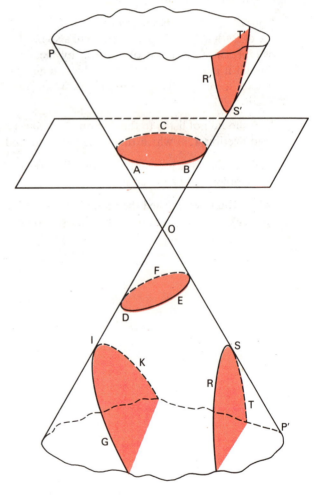

*Figure 1    Two conical surfaces and the sections made by intersecting planes.*

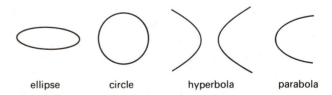

ellipse        circle        hyperbola        parabola

*Figure 2    The conic sections.*

6

# 1 INTRODUCTION

When problems can be modelled by shapes bounded by straight lines or circles, or by planes or cylinders, classical geometry, which was the central concern of Unit 4, is a form of description that works well, but it is useful only in situations where shapes are simple. If shapes are more complicated, classical geometry becomes less effective. A geometry based on graphical and algebraic methods is then needed.

You may remember from TV3, *Nothing New Under the Sun*, that an accurate description of the path of the Earth around the Sun was required in order to compare the time according to a sundial with Greenwich Mean Time. During the sixteenth and seventeenth centuries, the problems of planetary motion, as described by Copernicus and Kepler, indicated that more effective ways of describing the position of an object were needed. New ideas about the motion of bodies, culminating in Sir Isaac Newton's three laws of motion, required a study of the curves along which objects, principally projectiles for military purposes, would move. Problems of mapping the spherical Earth onto flat maps also called for closer knowledge of curves and their properties.

Against this background, Descartes and Fermat, both contributors to the discipline of algebra, saw the potential for algebraic solutions to problems in geometry. They were members of a French philosophical school which was interested in the structure and organization of human knowledge, and structure and organization was what they brought to geometry. They were looking for methods of solving geometrical problems that could be applied to all sorts of curves, whether complicated or simple, symmetrical or skew. They wanted systematic methods, rather than the more intuitive, symmetry-dependent methods used since the day of Euclid.

The foundations for some of the necessary developments were laid by the Greeks, who began the study of a set of curves called conic sections. You know the shape of an ice-cream cone: if I represent the shape of two cones placed with their pointed ends together in two dimensions, I obtain something like Figure 1. The conical surface consists of two parts, extending on opposite sides of the point of contact of the vertices, O, to an unlimited extent in both directions. I could take a knife and slice through this double cone in a number of different ways as Figure 1 shows. For example, the curve labelled ABC shows that the result of slicing horizontally through a cone is a circle. An inclined plane through just one of the cones gives an *ellipse* (DEF). In general, an ellipse or a circle will always result when a plane cuts through one part of the cone. If the cutting plane is inclined so as to cut both cones the resulting curve is called a *hyperbola* and has two parts, (R'S'T' and RST, for example). If, finally, the cutting plane is parallel to one of the lines of the cone (e.g. PP' or IOS'), the intersection is called a *parabola* (GIK, for example). The circle, ellipse, hyperbola and parabola will occur again later in the unit. Often they are referred to, not surprisingly, as the *conic sections*.

conic sections

These curves, which are useful in many modelling situations, are best described by equations rather than by geometrical shapes. The study that achieves such a description is called *co-ordinate geometry*.

## 1.1 Co-ordinate systems

Central to co-ordinate geometry is the use of a reference grid of the type used to draw a graph. In Unit 2, a reference grid was used, for example, to specify the variation of distance with time when modelling a train journey. Now, however, the grid is to be used to specify positions. For problems which deal with positions which lie on a plane, a flat surface, the system of co-ordinates is set up by selecting a reference position, called its *origin*, and then drawing two reference lines, each of which is a *co-ordinate axis* through the origin. Usually, the axes are drawn at right angles: the co-ordinate system is then called a *rectangular* co-ordinate system. This rectangular system is also called the Cartesian system after Descartes who devised it.

**Study comment**

**The co-ordinate systems discussed in this unit are all rectangular Cartesian co-ordinate systems. I shall therefore leave out the words 'Cartesian' and 'rectangular' when referring to co-ordinates, axes and so on, but you must remember that they are implied.**

A large scale example of a co-ordinate system like this is the Ordnance Survey National Grid, which provides a common system of reference for maps of the United Kingdom. Its origin lies to the south-west of the British Isles and the whole area is divided into 100 km major grid squares, designated by two grid letters (Figure 3).

Places listed in the handbook of the AA are referenced by dividing the major grid squares into 10 km squares. The 10 km square area in which a place of interest lies is then located by the grid letters and two numbers. The first number indicates the distance the south-west corner of the smaller area is to the east of the origin of the 100 km square: the second number indicates the distance north. Figure 4 shows the major grid square SP which contains Bletchley.

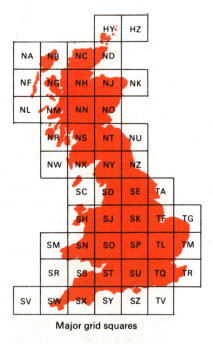

*Figure 3 The Ordnance Survey National Grid divides the British Isles into 100 km major grid squares, designated by two grid letters. Bletchley is located in the major grid square designated SP.*

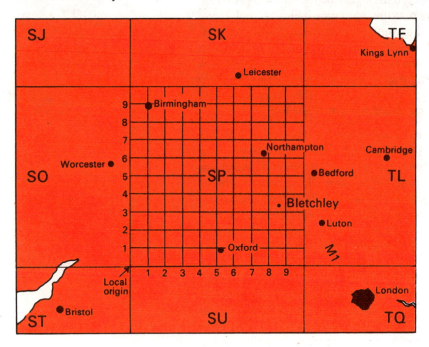

*Figure 4 The major grid squares are sub-divided into 10 km squares. The locations of places may be referenced by estimating their positions to the nearest kilometre.*

What is the reference for the area in which Bletchley lies and what do the two figures indicate?

The reference is SP83. This means that Bletchley lies somewhere in the 10 km square whose south-west corner is 80 km to the east and 30 km north of the origin of the major grid square designated SP.

To reference Bletchley more accurately, you could estimate the distances east and north from the origin of this smaller, 10 km square area, to the nearest kilometre say.

> What do you estimate the reference for Bletchley to be: (a) with respect to the origin of the 10 km square; and (b) with respect to the origin of the 100 km square, SP?

The scale of road maps allows you to find the position of a place with sufficient accuracy for driving from town to town, but many people have to use maps that are more accurate. The Ordnance Survey produce maps with a scale of 1:50 000, which are divided into 1 km squares. By estimating distances east and north in these squares, to a tenth of a kilometre (100 m), a six figure reference can be achieved. The first three figures, which are termed 'eastings', indicate the distance of the point of interest to the east of the origin of the major grid square and the second three figures, 'northings', give its distance north.* Using a 1:50 000 map, Bletchley station, by my estimation, has a grid reference, correct to 100 m, of SP(869,337).

(a) My estimate is that Bletchley is 6 km east and 3 km north of the origin of the 10 km square in which it lies and its reference is therefore (6,3) with respect to this origin.

(b) Using this estimate, Bletchley is 86 km east and 33 km north of the origin of the major grid square, SP. The reference in this case is (86, 33).

By using a co-ordinate system, the Ordnance Survey Grid, I first located a large area, then locate a place and finally I have located a point on a 1:50 000 map. These maps allow you to specify a position to the nearest 100 m (1/10 km). It should not be difficult to imagine a co-ordinate system that allows a position to be specified as precisely as may be wanted for any particular purpose. Figure 5 shows a co-ordinate system with its origin labelled O and the axes labelled X and Y. Conventionally, X is the label

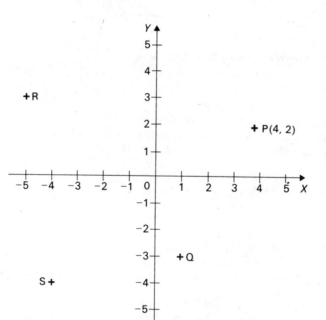

Figure 5 *A point is specified by, first, its distance to the right along the X-axis and then its distance up the Y-axis. Thus the point P is specified by its co-ordinates (4, 2).*

given to the *horizontal* axis and Y is used for the *vertical* axis. The location of a point, such as P, is described in this co-ordinate system by the distance from the origin to the right along the horizontal axis and the distance upwards from the origin along the vertical axis. In Figure 5 these intervals are shown as 4 and 2, respectively, and so that point P is described as having the *co-ordinates* (4, 2). The co-ordinates of the origin are (0, 0). All points that lie below the X-axis have negative values of Y and all points that lie to the left of the Y-axis have negative values of X.

> In this system, what are the co-ordinates of the points Q, R and S?

Q(1, −3); R(−5, 3); and S(−4, −4)

*If you are unfamiliar with the Ordnance Survey system try to find one or two locations on a 1:50 000 map yourself. An explanation of the reference system is given on each map.*

In all respects, the co-ordinate system for position I have just described is similar to that introduced in Unit 2 as the basis for plotting graphs of distance against time.

The Cartesian system can be extended to deal with problems that are three dimensional. There is no reason why a third reference direction, the Z-axis of Figure 6, should not be added to allow you to work with problems involving solid figures. The points A and B shown in this figure have co-ordinates (1, 3, 2) and (2, 2, 2), respectively, the Z-co-ordinate being written last in each case.

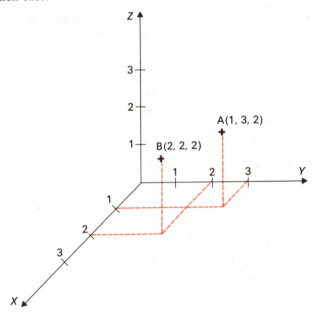

*Figure 6    A point may be specified in three dimensions by three co-ordinates (X, Y, Z) related to three mutually perpendicular axes.*

You probably know that Ordnance Survey maps not only allow you to estimate an easting and northing for a position, but also allow you to estimate its height by means of contour lines. A contour line is drawn through points that are at the same distance above or below sea level.

Imagine a co-ordinate for height that relates to a third axis which rises vertically out of the map from the origin of the Ordnance Survey Grid.* This co-ordinate is similar to the Z-co-ordinate in the three-dimensional system described earlier.

In using X, Y and Z labels for axes in three dimensions, it is usual to reserve Z for height so that X and Y refer to horizontal axes. A negative Z-co-ordinate then refers to depth below the X and Y axes.

Once a system of reference is defined by its origin and axes, the positions of points can be specified by their co-ordinates. Equations can then be set up that describe the relationships between points that lie, say, on the same straight line. By using algebra, these equations can be manipulated to calculate, for example, the distances between points. The algebra involved is similar whether you work in two or three dimensions, but for problems in three-dimensions it is more cumbersome and therefore I shall not pursue three-dimensional ideas any further in this unit.

*At present on Ordnance Survey 1:50 000 maps, the difference in height, or vertical interval, between one contour line and the next is 50 feet. The vertical interval is therefore approximately 15 metres. Heights on the contour lines are, however, given in metres.*

## 1.2   Distance and direction

In co-ordinate geometry you normally know the positions of a number of points, including the origin; the kinds of question you may be asked about these points are:

1   Are they at equal distances from one another?

2   Do they lie on the same straight line?

3   Do any of the points lie on a particular curve?

Alternatively, you may want to specify the co-ordinates of a point that is related to the known points.

To solve problems of this type, you can achieve much by the methods of classical geometry, but normally the problem cannot be treated completely by this approach. Let me be specific: points A, B and C have the co-ordinates $(1, 1)$, $(2, 2)$ and $(0, 4)$. Suppose you were asked if ABC is a right-angled triangle; or, in other words, whether the angle $A\hat{B}C$ is 90°. From Unit 4, you know that in a right-angled triangle: the sum of the two angles, other than its right-angle, is 90°; the triangle can be divided into two right-angled triangles which are equiangular with the given triangle; the sum of the squares of two shorter sides equals the square of the hypotenuse (Pythagoras' theorem). Using trigonometry, you also know the various ratios of the sides of the triangle for given angles.

> How can you use this information to prove that ABC is a right-angled triangle?

If you could calculate the lengths of the sides of the triangle, you could use the converse of Pythagoras' theorem. So first, I must show you how to calculate distances between points using co-ordinate geometry.

**The distance between two given points**

Figure 7 shows the points ABC for the problem given. The point D is a point I have constructed by drawing the line BD parallel to the $Y$-axis and the line AD parallel to the $X$ axis.

> Write down the co-ordinates of D.
>
> D(2, 1)

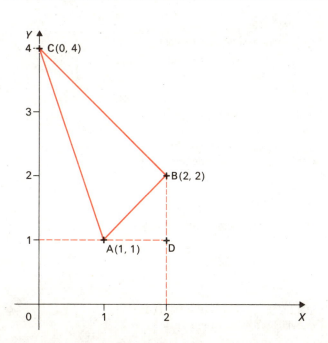

Figure 7   The triangle ABC is defined by the co-ordinates of its vertices A, B and C.

11

To find the length of AB, it is enough to note that the construction makes BD̂A a right-angle. The length of AD is the difference between the values of the $X$-co-ordinate of D and the $X$-co-ordinate of A, that is, $2 - 1 = 1$. Similarly, the length of BD is the difference between the $Y$-co-ordinates of B and D: $2 - 1 = 1$.

Using Pythagoras' theorem, for triangle ABD

$$(AB)^2 = (AD)^2 + (BD)^2$$
$$= 1^2 + 1^2$$
$$= 2$$

SAQ 1

**SAQ 1**

    (a)  Using a similar method, find the squared lengths of the lines BC and AC in the triangle ABC of Figure 7.

    (b)  Does Pythagoras' theorem apply to triangle ABC?

From SAQ 1, you can see that in this case $(AB)^2 + (BC)^2 = (AC)^2$ and so the triangle ABC is right-angled, with $A\hat{B}C = 90°$.

Notice here that I used Pythagoras' theorem to find the squared lengths of the sides of the triangle and its *converse* (see Unit 4) to establish that $A\hat{B}C = 90°$.

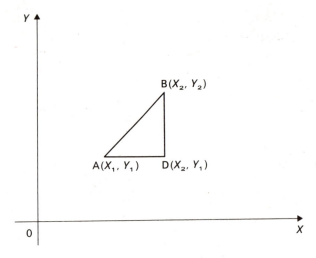

*Figure 8*  *The distance between two points A and B can be found from the equation* $(AB)^2 = (X_2 - X_1)^2 + (Y_2 - Y_1)^2$.

Having shown you the method of finding the distance between two points, I can work out a more general formula by choosing two arbitrary points A and B. If I denote their co-ordinates (Figure 8), by $(X_1, Y_1)$ and $(X_2, Y_2)$ the co-ordinates of the construction point D will be $(X_2, Y_1)$ and the length of AB can be found from

$$(AB)^2 = (DA)^2 + (BD)^2 = (X_2 - X_1)^2 + (Y_2 - Y_1)^2$$

Thus you now have a general formula for finding the distance between any two points whose co-ordinates are known.

SAQ 2

**SAQ 2**

    Figure 9 shows a triangle PQR with P(0, 2), Q(−3, −1) and R(3, −1). Is PQR a right-angled triangle, and if so, which angle is a right-angle?

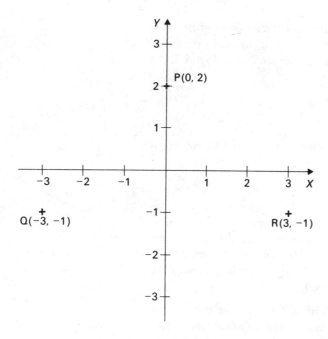

*Figure 9   See SAQ 2.*

### Gradients

You encountered the idea of a gradient when graphs were discussed. In that case, gradient was a quantity which indicates how steep a line is. In co-ordinate geometry the same term is used and it has the same meaning.

Consider the triangle ABC of Figure 7. What is the gradient of line AB?

From your earlier work, you already know this to be $BD/DA$. This gradient has the value $(2 - 1)/(2 - 1) = 1$. More generally, I want to define the gradient of the line AB in Figure 8. This is

$$M = \frac{\text{the difference in the } Y\text{-co-ordinates}}{\text{the difference in the } X\text{-co-ordinates}}$$

$$= \frac{Y_2 - Y_1}{X_2 - X_1}$$

Note that $M$ is the symbol that is used for gradients. What I have done is to start with the co-ordinates of the point at one end of the line, the point B, and subtracted from this the co-ordinates of the point at the other end of the line, the point A. As a specific example, consider the gradient of the line BC in Figure 7. For this line the gradient is

$$M = \frac{4 - 2}{0 - 2} = -1$$

I have put the co-ordinate values of C first in both numerator *and* denominator.

What would be the result if co-ordinate values of B are placed first?

$$M = \frac{2 - 4}{2 - 0} = -1 \text{ exactly as before.}$$

*So as long as the same order is adopted in the numerator and the denominator, the order of co-ordinates is not important.*

What would be the result of not adopting the same order in numerator and denominator?

13

The incorrect result of $+1$ would be obtained for the gradient of BC. It is incorrect because any line which slopes downwards from left to right in a graph is said to have a *negative* gradient. So you can see that, in Figure 7, the slope of BC (or CB) is negative.

### SAQ 3

What are the gradients of RS and ST if the points R, S and T have the co-ordinates $(3, 2)$ $(0, 1)$ and $(-1, 4)$ respectively? Is RST a right-angled triangle?

A geometrical gradient is used to measure the gradient of a railway line, where it is quoted as the height gained in a given horizontal distance. In this case, it is the tangent of the angle of the slope. When quoting the gradient of a road, however, it is often the rule to quote the ratio of height gained by the road to the distance travelled *along* the road.

This is not quite the same as the definition of the gradient used in co-ordinate geometry. However, for most roads the difference between the two definitions is very small; it is the difference between the *sine* and the *tangent* of the angle of the slope (Figure 10).

*Parallel and perpendicular lines*

Two questions about lines that can be answered by a knowledge of their gradients are: (i) are they parallel; (ii) are they at right angles?

Any line that has a gradient of $-1$ will be parallel to the line BC in Figure 7. Similarly, any line that has the same gradient as AB will be parallel to AB and any line that has the same gradient as AC will be parallel to AC.

### SAQ 4

If two new points are defined in Figure 7 by multiplying the co-ordinates of A and B by a factor $K$, will the line which passes through the two new points be parallel to any of the sides of triangle ABC?

For the points in Figure 7, the gradient of AB is simply the negative of the gradient of BC. The gradient of AB is $+1$, while that of BC is $-1$ and AB and BC are at right-angles. The answer to SAQ 3 also shows two lines that are at right-angles. These are RS (gradient $1/3$) and ST (gradient $-3$). The product of the gradients of AB and BC is

$$(+1)(-1) = -1$$

The product of the gradients of RS and ST is

$$(\tfrac{1}{3})(-3) = -1$$

So it seems that the product of the gradients of two lines that are perpendicular (which means at right-angles or $90°$ to each other) may always have the value $-1$.

I can justify the general truth of this by referring you to Figure 11. The line OA makes an angle $\theta$ with the $X$-axis: the gradient of OA is $\tan \theta$. I have constructed OB to be at right-angles to OA. Therefore $B\hat{O}C = (90° - \theta)$ and, since $O\hat{C}B$ is constructed as a right-angle, $O\hat{B}C = \theta$.

The gradient of OB is $-(BC/OC)$, since the line slopes in a negative sense, but this is $-(1/\tan\theta)$.

Therefore, the product of the gradients of OA and OB is

$$\tan\theta \times -\frac{1}{\tan\theta} = -1$$

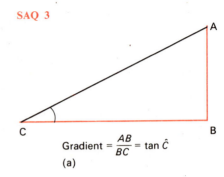

Gradient $= \dfrac{AB}{BC} = \tan\hat{C}$

(a)

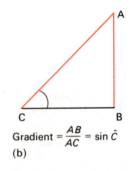

Gradient $= \dfrac{AB}{AC} = \sin\hat{C}$

(b)

*Figure 10 (a) A geometrical gradient is used to measure the slope of a railway track. (b) The gradient of a road, however, is only approximately equal to this.*

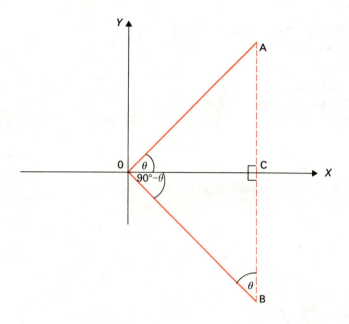

Since any line parallel to OA has the same gradient as OA and any line parallel to OB has the same gradient as OB, it follows that any pair of lines parallel respectively to OA and OB will also be perpendicular. Thus the product of the gradients of any pair of perpendicular lines is − 1.

*Collinearity*

Another question that can be answered by considering just the gradients of lines is whether three points lie in a straight line, that is, whether they are *collinear*.

What I need to do is show that the line from A to C (Figure 12(a)) has the same gradient as the line from A to B. If this is true, the construction of the line AC must take it through the point B.

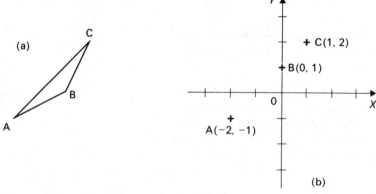

Figure 12 *If the gradient of AC is the same as the gradient of AB, then AC and AB are parallel and B must lie on the line AC.*

Consider the points A, B, C in Figure 12(b). They have co-ordinates (− 2, − 1), (0, 1), and (1, 2). They look as if they lie in a straight line, but can we show that they lie in a line using the methods of co-ordinate geometry? Since I have just derived a formula for gradients, perhaps knowledge of this can help me. The gradient of AC is

$$M_1 = \frac{2-(-1)}{1-(-2)} = \frac{2+1}{1+2} = 1$$

The gradient of AB is

$$M_2 = \frac{1-(-1)}{0-(-2)} = \frac{1+1}{0+2} = 1$$

15

So the gradients $M_1$ and $M_2$ are equal. Since both lines pass through the point A and have the same gradient the points A, B, and C must lie in a straight line.

Equally, I could have shown that AC and BC are parallel or that AB and BC are parallel. If two lines are parallel and they pass through a common point, then all points on the lines must be collinear.

**SAQ 5**

Do the points P(−4, 3), Q(−2, −1), R(−1, −3) shown in Figure 13 lie on a straight line?

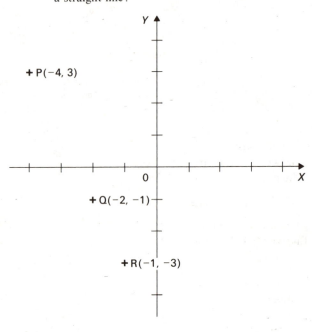

*Figure 13    See SAQ 5.*

### 1.3    The importance of being straight

In Unit 2 you met several equations for various straight lines. I want to look at such equations again, but this time I am interested in finding equations that describe lines which do specific things.

Since it is not possible to draw more than one straight line through two fixed points, any straight line is fixed by knowing the co-ordinates of any two points along its length. This means that given two points on a line it should be possible to predict every other point through which the line passes. This can be done by setting up an equation for the line that passes through the two known points.

From an ability to calculate a gradient, it is a short step to derive an equation for any straight line. After all, what distinguishes a straight line is the constancy of its direction. In the terms of co-ordinate geometry, a straight line is one whose inclination is fixed relative to the reference axes. It is a line of constant gradient.

How, for example, do I arrange for an equation to describe a line through the two points A and B? Is there any special feature about the equations for *all* the lines which pass through the single point A or through the single point B?

**The general equation of a straight line**

Some of the simpler equations for straight lines given in Unit 2 had the form $Y = MX$. Is $M$ equal to the gradient of a line in an equation of this form?

Let me substitute some values for $X$.

When $X = 0$, then $Y = 0$

When $X = 1$, then $Y = M$

When $X = 2$, then $Y = 2M$

Two points on the line are therefore $(1, M)$ and $(2, 2M)$. Using the method outlined in Section 1.2, I can write the gradient of the line as

$$\frac{2M - M}{2 - 1}$$

which reduces to $M$, so $M$ *is* the gradient.

This line $Y = MX$ passes through the origin; some lines do not, so the *general equation* for a straight line is

$$Y = MX + C \tag{1}$$

where $M$ is the gradient and $C$ is the value of $Y$ when $X = 0$. The value of $C$ is called the *intercept* of the line (with the $Y$-axis). In Figure 7, the intercept of BC with the $Y$-axis is 4. When $C = 0$ the line passes through the origin, since then $Y = 0$ when $X = 0$. Both OA and OB in Figure 11 have zero intercepts. intercept

### SAQ 6 SAQ 6

In Unit 3, Section 2, a particular linear relationship is found to be $P = N + AY$, where $A$ and $N$ are parameters and $P$, $Y$ are variables.

This should be the equation of a straight line. What are the gradient and intercept (with the $P$-axis) of this line? Where does it cross the $Y$-axis?

**The equation of a line passing through a fixed point and of known gradient.**

A straight line has constant gradient, by definition, and this idea will help to give the line an equation.

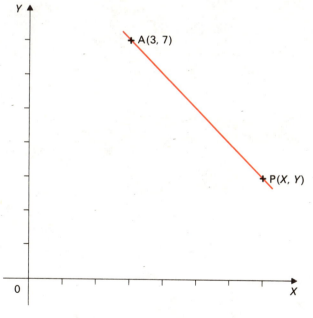

*Figure 14    The gradient of the line AP is given by $-1.5 = (Y - 7)/(X - 3)$.*

Consider the straight line passing through the point A(3, 7) as shown in Figure 14. What I want to find is a relationship between this fixed point A and any other point on the line. Suppose I choose an arbitrary point P(X, Y) that is on the line. I can now find an expression for the gradient of this line by using the method of subtracting co-ordinates outlined in Section

17

1.2 The gradient is

$$\frac{Y-7}{X-3}$$

and this is true whatever the actual position of P, as long as it lies on the line, but does not coincide with A. Suppose I know the gradient of the line is, say, $-1.5$. Since this is constant for this line, then for all possible positions of P

$$\frac{Y-7}{X-3} = -1.5. \tag{2}$$

So this is the equation for the line through A(3, 7) of gradient $-1.5$. You may not recognize it in this form, but it *is* one since I need only specify the value of $X$ in order to calculate the corresponding value of $Y$. For example, if I put in the value $X = 5$, then I can deduce that

$$\frac{Y-7}{2} = -1.5.$$

so that $Y - 7 = -3$, or $Y = 4$.

I can rearrange equation (2) by multiplying both sides by $(X - 3)$, thus

$$Y - 7 = (X - 3)(-1.5)$$

After multiplying out the brackets, this becomes

$$Y = -1.5X + 4.5 + 7$$

or

$$Y = -1.5X + 11.5.$$

This is the *standard form* of the equation of a line.

What if the gradient is $M_1$ rather than $-1.5$ and the line passes through some fixed point A($X_1$, $Y_1$)? Let P($X$, $Y$) be a point on the line that has a gradient of $M_1$. Then the line AP is fixed and its gradient has the value $M_1$. The gradient of line AP is therefore given by

$$\frac{Y - Y_1}{X - X_1} = M_1 \tag{3}$$

or

$$(Y - Y_1) = M_1(X - X_1) \tag{4}$$

This is the equation for a line through the point A($X_1$, $Y_1$) with gradient $M_1$.

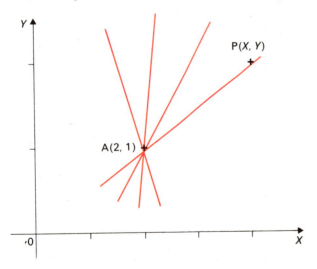

*Figure 15    All the lines through the point A (2, 1) have equations of the form* $(Y - 1) = M(X - 2)$.

Figure 15 shows a number of possible straight lines all passing through the point A(2, 1). Lines through A(2, 1) must have equations of the form $(Y - 1) = M(X - 2)$. For lines with different gradients it is necessary only to change the value of $M$, since only the gradients can be different.

18

## SAQ 7

Find the equation of the line passing through the point E(0, $W$) and having gradient $M$ and write the equation in standard form.

## SAQ 8

Find an expression for the equation of a line through the point B($-2, -5$) with a gradient $M$. Substitute the value $M = -1$ and put the equation in standard form.

### The equations for a line passing through two fixed points

If I know that a line passes through two fixed points, I can again use the idea of a constant gradient to find the equation of the line. Let me repeat the problem for given arbitrary points A($X_1, Y_1$), B($X_2, Y_2$). Now the gradient is

$$M = \frac{Y_2 - Y_1}{X_2 - X_1}$$

and this is fixed since A and B are fixed.

The equation may be set up by writing the gradient between A($X_1, Y_1$) and any other point P($X, Y$) as

$$M = \frac{Y - Y_1}{X - X_1}$$

These values of $M$ must be the same, since $M$ is the gradient of the line.

$$\frac{Y - Y_1}{X - X_1} = \frac{Y_2 - Y_1}{X_2 - X_1} \tag{5}$$

The variation of $X$ and $Y$ alone will describe the line since the other terms are fixed. This is the equation of the straight line through any two fixed points.

## SAQ 9

Find the equation of the line joining the points C(2, 4) and D(4, $-3$) and write it in standard form

### Equations for parallel lines

I am now able to discuss the relationship between the equations of parallel lines in more detail.

Figure 16 shows a series of parallel lines. All can be described by the general equation of a straight line

$$Y = MX + C$$

The fact that the lines are parallel means in co-ordinate geometry that their gradients are equal. However, it is not possible for them to have the same intercept as can clearly be seen from the figure. There is a relationship between the equations of these lines because they all have the same value of $M$ and different values of $C$.

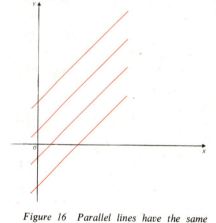

*Figure 16  Parallel lines have the same gradient, but they cut the Y-axis at different points, that is, they have different intercepts.*

## SAQ 10

Refer back to Figure 9 and find the equation of line PR without making use of the co-ordinates of R. (Hint: you should be able to find the equation from one point and a gradient. Since PQR is a right-angled triangle you should be able to deduce the gradient of PR from that of PQ.)

19

## 1.4 Summary

In summary there are only two parameters, $M$ and $C$, known as the *gradient* and the *intercept* (with the $Y$-axis) in the general equation of a straight line, $Y = MX + C$, so only two conditions are needed to fix a line.

A line is specified either by knowledge of its *gradient* and *one point* on the line, or by knowledge of *two points* on the line.

Parallel lines have the same gradient. The product of the gradients of perpendicular lines is $-1$. Note that this means that

$$\text{one gradient} = \frac{-1}{\text{the other gradient}}$$

# 2 CIRCLES

**Study comment**

**Before reading this section it would be helpful for you to read or re-read Town Planning in *Modelling Themes***

## 2.1 Circles and ground rent

Let me recap on a problem that you met in Unit 2, in connection with the intersection of straight lines. I will reiterate and develop the details of this problem in order to show how the circle can be a useful form of description.

A farmer may choose between crops according to the net profit per unit yield. The problem is to decide where, with respect to the town centre, it will be possible to pay a certain land rent and still make a profit (when growing a certain crop). Von Thunen's model supposes that the profitability depends only upon the crop yield, the cost of producing each unit of this crop and the generalized cost of transporting the crop to the town centre.

According to this simple model, the result of plotting the highest rent that a farmer would be prepared to pay against the distance from the town centre would be a different straight line for each crop. Consider the situation for two different crops, as in Figure 17. I have labelled the intersection of the maximum rent–distance lines as P. This represents the point at which a

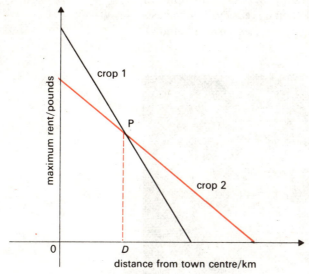

*Figure 17    A graphical comparison of Von Thunen's agricultural model for two crops. The point P indicates where one crop becomes more profitable to grow than the other. When a crop is 'more profitable' a higher maximum rent can be paid.*

changeover in crops is likely. The reason for this changeover is the desire of the farmer always to grow a crop that is profitable in a situation where the landowner will demand the highest rent he can. The model leaves out many variables that might affect the situation—the quality of land drainage, the availability of roads and vehicles. It is a simplification of reality. Let us suppose that on the whole, rents will be highest near the town centre where competition for land is at its most fierce. Up to a distance $D$ km from the town centre you are likely to find mainly crop 1 being grown: beyond this, crop 2 becomes more profitable.

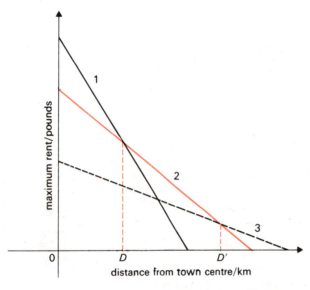

Figure 18 *D and D' are the distances from the town centre where crop 2 becomes more profitable than crop 1 and where crop 3 becomes more profitable than crop 2.*

It is straightforward to extend this idea to a third crop as in Figure 18. Crop 1 will still outbid the others up to a distance $D$ km from the town centre. Beyond this, and up to a greater distance $D'$ km, crop 2 will flourish. Beyond $D'$ km crop 3 will tend to be grown more.

According to the model, if you were to take an aerial view of the resulting situation, you should be able to see the pattern indicated in Figure 19. The fact that a certain distance in any direction from the town centre determines a change in land use implies a circular pattern of development with $D$ and $D'$ as the radii of two of the circles. Indeed, if you suppose that a similar model is valid for non-agricultural land uses as well, then it is easy to see the basis for the concentric ring model of urban development (see Town Planning).

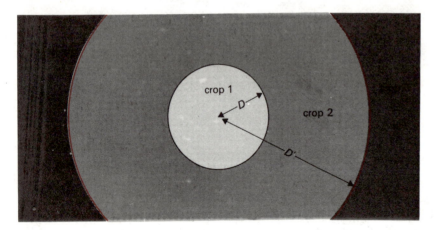

Figure 19 *A circular pattern of land usage is predicted by Von Thunen's model.*

Circles are among the simplest of the curved figures for geometrical models. How can circles be described using the techniques of co-ordinate geometry?

## 2.2 An equation for the circle

I shall take a single circular boundary and begin with the simplest possible case—that in which the town centre or market place is the origin. Suppose the boundary has a radius $A$ and I choose a point $P(X, Y)$ on the boundary

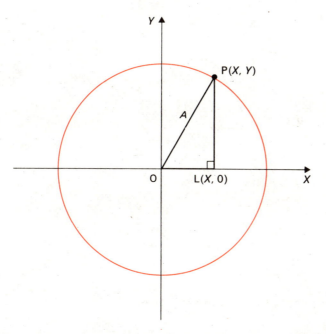

*Figure 20    A circle is specified by its radius and the co-ordinates of its centre. The construction required to show that the equation of a circle centred at the origin $(0,0)$ and of radius $A$ is $A^2 = X^2 + Y^2$.*

as in Figure 20. If the circle can be described by an equation, the values of $X$ and $Y$, which are the co-ordinates of the point P, must be related to one another. It is this relationship that I must look for. Notice that I have chosen the point P to lie on the boundary in the first quadrant, which gives only positive values of $X$ and $Y$. Regardless of its position, however, I could make the same construction.

I can construct a right-angled triangle OPL by placing L on the X-axis so that the angle $\hat{PLO}$ is 90° and the co-ordinates of L are $(X, 0)$. Also, because P lies on the circle, the length OP is equal to $A$, the radius of the circle, while the length of PL is the $Y$-co-ordinate of the point P.

Applying Pythagoras' theorem to the triangle OPL

$$(OP)^2 = (LO)^2 + (LP)^2$$

or

$$A^2 = X^2 + Y^2$$

This last equation relates the $X$ and $Y$-co-ordinates of the point P and it must be true wherever P may lie on the circle. The equation is thus the general equation of a circle of radius $A$ centred on the origin. The form of this equation is such that for every value of $X$ there are two values of $Y$ (positive and negative). For example, where $X = 0$, $Y = +A$ and $-A$. Usually, this is written $X = 0$: $Y = \pm A$. Notice that the number which multiplies $X^2$ is equal to the number which multiplies $Y^2$. You also know that there is no real number which has a negative square and $A^2$ must be positive for this to be the equation of a circle.

### SAQ 11

SAQ 11

What curve is represented by: (a) $X^2 + Y^2 = 16$; and (b) $9X^2 + 9Y^2 = 1$. If they are circles, what are their radii?

The equations in SAQ 11 are restricted to circles centred on the origin, but suppose I were concerned with an agricultural location problem where the market and the town centre do not coincide. Fortunately, to deal with the more general case of a circle centred on the point $Q(X_1, Y_1)$ the same sort of calculation can be used. Figure 21 shows the construction. This time I shall

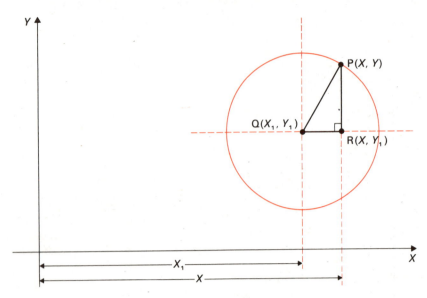

*Figure 21    If the centre of a circle of radius A is at the point $(X_1, Y_1)$, its equation is $X^2 + Y^2 - 2X_1 X - 2Y_1 Y = A^2 - X_1{}^2 - Y_1{}^2$.*

specify a point R which lies on a line drawn through Q parallel to the $X$-axis. R now has the co-ordinates $(X, Y_1)$ and the length of PR is $(Y - Y_1)$, the length of QR is $(X - X_1)$ and the length of QP is, as before, the radius $A$.

Applying Pythagoras' theorem, as before, gives

$$(QP)^2 = (QR)^2 + (PR)^2$$

or

$$A^2 = (X - X_1)^2 + (Y - Y_1)^2$$

This is the required equation of a circle of radius A, centred on the point $(X_1, Y_1)$. Note that the effect of moving the centre of the circle to $(X_1, Y_1)$ is to replace $X$ by $(X - X_1)$ and $Y$ by $(Y - Y_1)$.

The bracketed terms can be multiplied out: this makes the equation look more formidable, but it still represents a circle. You will remember from Unit 4, that generally

$$(A - B)^2 = A^2 - 2AB + B^2$$

Now the equation of the circle is

$$A^2 = (X - X_1)^2 + (Y - Y_1)^2$$

I can replace each of the squared brackets by their expanded versions to give

$$A^2 = (X^2 - 2XX_1 + X_1^2) + (Y^2 - 2YY_1 + Y_1^2)$$

and arrange this in a similar order to the general equation for a circle centred at the origin.

$$X^2 + Y^2 - 2XX_1 - 2YY_1 = A^2 - X_1^2 - Y_1^2$$

The point of showing you this general equation is that you will not always meet the equation of a circle in a form that is immediately recognizable.

Let us examine the equation

$$X^2 - 12X + Y^2 - 4Y + 31 = 0$$

I want to know if it represents a circle and, if it does, I want to know the centre and radius of that circle. It contains terms in $X^2$ and $Y^2$ and terms which are multiples of $X$ and $Y$, and a constant, so it may represent a circle.

As before, there are two requirements for the equation to be the equation of a circle. The first requirement, that the number of $X^2$s should be equal to the number of $Y^2$s, is satisfied by the equation.

You can see that if by equating the constant in a given equation to the constant term in the general equation for a circle the value of $A^2$ turns out to be negative, then $A$ does not have any real value (remember that the square root of a negative number is not possible) and the given equation would not be that of a circle. Thus the second condition for an equation of a circle is that $A^2$ is positive.

In order to see if the equation given represents a circle, compare it with the general equation of a circle, thus

$$X^2 + Y^2 - 12X - 4Y = -31$$
$$X^2 + Y^2 - 2X_1 X - 2Y_1 Y = A^2 - X_1^2 - Y_1^2$$

You will note that I have re-arranged the equation of interest to look like the general equation of a circle as much as possible. If the equation of interest *is* that of a circle then it should be possible to make the two equations look *identical*. Certainly $(X^2 + Y^2)$ appears in both equations. For $-2X_1 X$ in the general equation, $-12X$ appears in the equation of interest. These two would be the same if

$$-2X_1 = -12$$

or

$$X_1 = 6$$

The terms in $Y$ would be the same if

$$-2Y_1 = -4$$

or

$$Y_1 = 2$$

Finally the *constant* terms (terms containing neither of the variables $X$ and $Y$) would be the same if

$$A^2 - X_1^2 - Y_1^2 = -31$$

I can substitute the values of $X_1$ and $Y_1$ that seem to fit, so

$$A^2 - 6^2 - 2^2 = -31$$
$$A^2 - 36 - 4 = -31$$
$$A^2 - 40 = -31$$

so

$$A^2 = 9$$
$$A = \pm 3$$

$A$ is the radius of the circle, so the minus sign has no significance: you can measure the radius in either direction. If $A^2$ had been negative, then $A$ would not have had any real meaning and the equation would not have been that of a circle.

Since both requirements are satisfied the given equation *is* that of a circle with its centre at $(6, 2)$ and a radius of 3.

The multipliers of $X$ and $Y$ are called the coefficients of $X$ and $Y$ and one of the processes I have just gone through is called *equating coefficients*. When I put $-2X_1 = -12$, I was equating coefficients of $X$ in each equation, and when I put $-2Y_1 = -4$, I was equating coefficients of $Y$. Equating $A^2 - X_1^2 - Y_1^2$ and $-31$ is called *equating constant terms*.

> What would I have done if I had been presented with an equation that contained $9X^2 + 9Y^2$ as well as multiples of $X$ and $Y$ and a constant?

This equation still satisfies the first requirement for it to be the equation of a circle: the number of $X^2$s and the number of $Y^2$s are equal. But comparison with the general equation will not work unless I divide throughout by nine.

Note that this was the technique required for the second equation in SAQ 11.

## SAQ 12

The equation $X^2 - 8X + Y^2 - 6Y = 0$ represents a circle. Find its radius and centre.

## SAQ 13

Write down an equation, in terms of the co-ordinates $(X_1, Y_1)$ of its centre, to which the circle of radius 5 must conform.

## SAQ 14

Farmers in the vicinity of a certain town may choose between two types of crop. Crop A yields 10 000 kilograms per square kilometre and crop B yields 20 000 kilograms per square kilometre. Crop A is expected to give £0.05 per kilogram (gross) profit when sold in the town centre, whereas Crop B will give only £0.03 per kilogram profit. The cost of transporting both crops to the town centre is £0.001 per kilogram per kilometre.

(a) Determine the distance from the town centre at which the 'changeover' between the two crops occurs.

(b) Would you expect to find crop A being grown at 7 km from the town centre?

(c) If the main market is transferred to a point 5 km east and 10 km north of the town centre, but the transport cost per unit mass per unit distance remains the same, write down an equation for the expected 'changeover' boundary between crops 1 and 2 (keep the town centre as origin and use a distance grid based on kilometre squares, the $Y$-axis running SN and the $X$-axis running WE).

(Reminder: Von Thunen's agricultural model states

$$R = EP - ETD$$

where:

$R$ is the number of pounds the producer can pay in rent per square kilometre;

$E$ is the number of kilograms of crop the land yields per square kilometre;

$P$ is the number of pounds (gross) profit per kilogram of crop;

$T$ is the number of pounds transport costs per kilometre and per kilogram of crop; and

$D$ is the number of kilometres the land is from the town centre.)

## 2.3 Shopping in circles

One of the lessons to be learnt from analysis of shopping data (and perhaps from common sense) is that the number of journeys to a particular shopping centre from a residential zone will depend upon the distance between the residential zone and the shopping centre. Indeed, this number declines with increasing distance. The question is whether or not it is valid to suppose that the number of journeys declines with distance in a systematic way according to a fixed curve. Another question is whether or not the number of journeys varies with distance in the same way between every shopping centre and residential zone. Suppose that the answer to both questions is yes.

What lines or curves that you have met so far in the course could represent the decline in the number of journeys with distance?

A plot of number of journeys against distance between residential zone and shopping centre may result in a straight line with a negative gradient. In fact, this does not fit with the available data, which show quite a steep fall-off with distance at first, followed by a more gradual decline.

Could an exponential curve of the sort that you met at the end of Unit 3 help?

This possibility might not occur to you, since in Unit 3 you will have associated such curves with growth, but look a little more closely at the equation of an exponential curve. I can write it in the form

$$Y = A^X$$

where $A$ is a constant which is larger than 1, and $X$ and $Y$ are variables. As $X$ increases, so does $Y$, but at an increasingly rapid rate. If $A = 2$, Table 1 shows how $Y$ increases with $X$ for $X = 1$ up to $X = 5$.

**Table 1  Values of $X$ and $Y$ for $Y = 2^X$**

| $X$ | 0 | 1 | 2 | 3 | 4 | 5 |
|---|---|---|---|---|---|---|
| $Y$ | 1 | 2 | 4 | 8 | 16 | 32 |

What would happen if I made the exponent negative? In other words, how does $Y$ vary if I write $Y = A^{-X}$?

Take $A = 2$ again and remember that $A^{-2}$ means $1/A^2$: you will find that $Y$ varies as in Table 2.

**Table 2  Values of $X$ and $Y$ for $Y = 2^{-X}$**

| $X$ | 0 | 1 | 2 | 3 | 4 | 5 |
|---|---|---|---|---|---|---|
| $2^X$ | 1 | 2 | 4 | 8 | 16 | 32 |
| $2^{-X} = Y$ | 1 | 0.5 | 0.25 | 0.125 | 0.0625 | 0.03125 |

Therefore $Y$ decreases very rapidly at first and then more and more slowly. This means that a negative exponential curve might also fit shopping data.*

A similar model is one in which the exponent is not one of the variables. What sort of curve would I get if its equation were $Y = X^{-A}$?

Again I will let $A = 2$ and draw up a table of values of $X$ and $Y$ for $X$ varying between 1 and 5. Note that with $A = 2$ the equation reads $Y = X^{-2}$, which means $Y = 1/X^2$.

**Table 3  Values of $X$ and $Y$ for $Y = X^{-2}$**

| $X$ | 1 | 2 | 3 | 4 | 5 |
|---|---|---|---|---|---|
| $X^2$ | 1 | 4 | 9 | 16 | 25 |
| $X^{-2} = Y$ | 1 | 0.25 | 0.11 | 0.0625 | 0.04 |

Again this is the right sort of variation for shopping data. The equation $Y = 1/X^2$ is sometimes called an *inverse square* relationship, since it involves 'one-over-something' raised to the power of two. Note also that I cannot put $X = 0$ in this equation without causing distress. One divided by zero is undefined, it is not a number that I could plot on a graph; so this is not an equation that should be used if $X$ stands for something that could be zero (or very near to zero, in which case $Y$ would be uncomfortably large— try $X = 0.01$).

*This possibility will be followed up in TV5.*

In the context of modelling shopping journeys, I can use $Y$ to represent the number of journeys and $X$ to represent distance. More precisely, I can use $X$ to represent the cost of travel by a particular mode of transport from the residential zone to the shopping centre, which is related to distance (see Town Planning). If all the distances of interest were to be small, I would not happily use an inverse square relationship. Nevertheless the inverse square variation is well-tried and used by town planners.

Suppose that I have a planning problem where lengths of shopping journeys with which I am concerned are large and where I am happy that I can use an inverse square relationship, since it seems to fit some existing data reasonably well. Faced with a population increase in certain areas of a town rather like Derby (see Town Planning, Figure 13), I have to decide upon sensible priorities with regard to shopping development. Specifically, I have to deal with shopping for food and groceries. There is a main central shopping area and there are out-of-town or local shopping areas. I will suppose that the number of journeys or trips, $T$, made to a particular centre depends upon its *attractiveness*, $F$, as well as its distance, $D$ m, from a particular residential zone. I will suppose also that the number of journeys should depend upon the amount of money, $£E$, that the occupants of a given residential zone will spend on food and groceries. In fact, this supposition suggests that the more money that people have to spend the more shopping journeys they will make.

How can I express these relationships mathematically?

The term 'proportional to', and the *proportionality* sign, $\propto$, are commonly used in mathematics. For example, I am supposing that if $F$ increases by a factor of two, then $T$ increases by a factor of two; if $F$ is halved, then $T$ will be reduced by a factor of two. So it appears that the number of journeys is *proportional* to the attractiveness of the centre, and I write

*proportionality*

$$T \propto F$$

Since $T$ also depends on $E$ and $D$, the statement is true *only* if $E$ and $D$ do not alter. I can also write

$$T \propto E \qquad \text{(if } F \text{ and } D \text{ are constant)}$$

and using the inverse square model I can write

$$T \propto \frac{1}{D^2} \qquad \text{(if } E \text{ and } F \text{ are constant)}$$

To convert these proportionalities into equations, I must introduce *constants of proportionality*.

For example, if $T$ depends on $F$ alone I can write

$$T = K_1 F$$

If $T$ depends on $E$ alone, I can write

$$T = K_2 E$$

If $T$ depends on $D$ alone, I can write

$$T = \frac{K_3}{D^2}$$

My suppositions are that $T$ depends on all three, that is, $E$, $D$ and $F$ simultaneously. If I also suppose that $F$, $E$ and $D$ are quite independent of each other then I can put all of these statements together as

$$T = \frac{KEF}{D^2}$$

where $K$ is the overall constant of proportionality (which differs from $K_1$, $K_2$ and $K_3$). You can check for yourself what this means if $E$, $D$ or $F$ alone is doubled while the other two are held constant. This type of argument is used frequently in setting up mathematical models. There are all sorts of factors that determine attractiveness as discussed in the Town Planning essay, but I will suppose that *floor area* is the sole factor. This is a common supposition by planners and you are perfectly entitled to doubt it, but I shall adopt it here for the sake of setting up a simple model (it will be discussed more fully in TV5).

It would be useful to identify the 'dragnet' spread by each shopping centre. In other words, I would like to know the boundary of choice between one shopping centre and another. On one side of the boundary, shoppers will tend to go to one centre more often; on the other side, they will tend to go most frequently to the other centre. So I will look simply at the 'conflict' for custom (total expenditure £$E$) between the main shopping centre with floor area $F$ m$^2$ and an out-of-town centre 4 km to the west with only half the floor area.

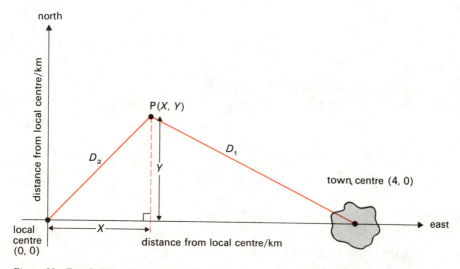

*Figure 22* *For shoppers living at P(X, Y), the local shopping centre at (0, 0) is equally as attractive as the town centre at (4, 0).*

This situation is represented in Figure 22 where I have constructed co-ordinate axes running SN and WE *with the local centre at the origin*, the town centre at (4, 0) and a residential zone centred at P($X$, $Y$) from which the occupants are likely to make just as many trips to the central shopping area as to the local one. All points like P define the boundary of choice that I want to find. Any residents nearer to the town centre than P are more likely to go to the central area than to the local one: all residents nearer the local centre than P are more likely to go to the local centre.

For the main shopping centre

$$T = \frac{KEF}{D_1^2}$$

where $D_1$ m is the distance from the residential zone at P to the town centre.

The attractiveness of the local centre is $F/2$ m$^2$, since it has half the floor area of the main centre. Thus for the local centre

$$T = \frac{KE(F/2)}{D_2^2}$$

$$= \frac{KEF}{2D_2^2}$$

where $D_2$ m is the distance from the residential zone at P to the local shopping centre. Since I have defined P as a point at which the numbers of trips to either centre are equal I can write

$$\frac{KEF}{D_1^2} = \frac{KEF}{2D_2^2}$$

29

Since $KEF$ appears on both sides of the equation

$$\frac{1}{D_1^2} = \frac{1}{2D_2^2}$$

and therefore

$$2D_2^2 = D_1^2$$

Using Pythagoras' theorem and the co-ordinates given in Figure 22

$$D_2^2 = X^2 + Y^2$$

and

$$D_1^2 = (4 - X)^2 + Y^2$$

The relationship between $D_2$ and $D_1$ can now be rewritten as

$$(4 - X)^2 + Y^2 = 2(X^2 + Y^2)$$

Multiplying out the brackets

$$16 - 8X + X^2 + Y^2 = 2X^2 + 2Y^2$$

Rearranging the terms in the equation gives

$$X^2 + Y^2 + 8X - 16 = 0$$

This is an equation of a curve which joins together all points such as P, where residents will find the two shopping centres equally attractive.

Is the curve a circle?

The first test, that the coefficients of $X^2$ and $Y^2$ should be equal, is passed immediately.

I can now compare the equation from the shopping model with the general equation of the circle centred on $(X_1, Y_1)$ and with radius $A$.

$$X^2 + Y^2 + 8X = 16$$
$$X^2 + Y^2 - 2XX_1 - 2YY_1 = A^2 - X_1^2 - Y_1^2$$

If I carry out the process of comparing coefficients, I can write down the following equations

$$-2X_1 = 8 \quad \text{or} \quad X_1 = -4$$
$$-2Y_1 = 0 \quad \text{or} \quad Y_1 = 0$$

The centre of the circle is therefore at $(-4, 0)$. By comparing the constants

$$A^2 - X_1^2 - Y_1^2 = 16$$

and substituting $X_1 = -4$ and $Y_1 = 0$

$$A^2 - 16 = 16$$
$$A^2 = 32$$
$$= 16 \times 2$$

Therefore

$$A = \sqrt{(16 \times 2)}$$
$$= \sqrt{16} \times \sqrt{2}$$
$$= 4\sqrt{2}$$
$$= 5.66 \text{ (to three significant figures)}$$

So the equation is that of a circle with its centre at $(-4, 0)$ and of radius 5.66.

This means that the boundary of choice must cross the $X$-axis when $X = -4 + 5.66$ and when $X = -4 - 5.66$, that is, at the points $(1.66, 0)$ and $(9.66, 0)$. I have drawn the complete curve in Figure 23.

I can give more justification for this result by asking you to think again of the model. The greater attractiveness of the town centre, its stronger pull,

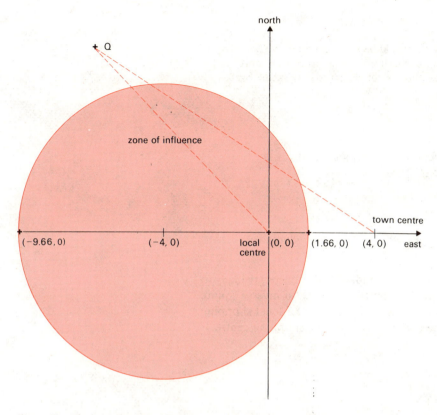

*Figure 23   The local shopping centre and the town centre are equally attractive to shoppers living on the 'boundary of choice', which in this case is a circle. In other words, the local centre has a circular 'zone of influence'.*

means that, between the two centres and anywhere on the right-hand side of the *Y*-axis, the point at which the number of trips will balance is likely to be nearer to the local centre than to the town centre: as I have found. If the two centres were to be equally attractive then the point of balance on the *X*-axis would be half-way between the two. On the left-hand side of the *Y*-axis at large distances from the local centre (such as the point Q marked in Figure 23) the greater attractiveness of the town centre means that a shopper is more likely to go there. The shopper's reasoning would be: 'I have to make a lengthy trip to either centre; the difference in travel costs will not be that large so, I might as well go to the more attractive centre'.

According to this model the local centre has a zone of influence bounded by the circle that I have drawn in Figure 23. Any population increase within this zone will tend to increase the trade in the local centre rather than that in the central area. This in turn might lead to an increase in the number of retailers wishing to move into the local centre, thus increasing the floor area and requiring the calculation to be done again. The model will be further complicated by the presence of many local centres and residential zones with differing total expenditures on food and groceries.

### SAQ 15

SAQ 15

Where would the boundary of choice lie if both centres shown in Figure 22 were equally attractive? What would be the equation of this boundary?

### SAQ 16

SAQ 16

Would the zone of influence of the local centre still have a circular boundary if the main (town centre) area were to be three times as attractive as the local centre (instead of twice as attractive)?

31

Keeping the same axes, what would be the equation of the boundary of influence of the local shopping centre if it were to be 4 km north of the position considered in the above example, but still the same size?

## 2.4  Summary

The equation for a circle with centre $(X_1, Y_1)$ and radius $A$ is

$$(X - X_1)^2 + (Y - Y_1)^2 = A^2$$

which can also be written

$$X^2 + Y^2 - 2X_1X - 2Y_1Y = A^2 - X_1^2 - Y_1^2$$

For a circle *centre at the origin* ($X_1 = 0$, $Y_1 = 0$) this reduces to

$$X^2 + Y^2 = A^2$$

Von Thunen's agricultural location model indicates that crops are likely to be grown in concentric circles centred on the market. An inverse square shopping model indicates that the 'zone of influence' of a local shopping centre may be a circle, not necessarily centred on the shopping centre.

# 3  THE OTHER CONIC SECTIONS

In the introduction to this unit, I referred to the ellipse, hyperbola and parabola as conic sections and pointed out that these curves first came to notice as part of the Greek study of geometry. They were then known as the curves which appeared when a cone was sectioned in certain ways (refer back to Figures 1 and 2). Later studies of the curves established that they had a number of interesting properties. Important points which relate to these curves, but do not lie on them are their *foci*. As the name given to these points suggests, reflecting surfaces, when bent into the form of conic sections, can be used to focus light, radio and sound waves.

**focus**

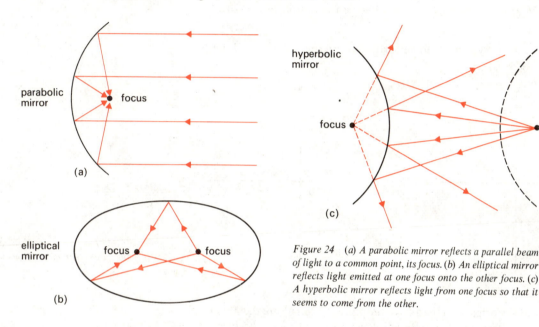

Figure 24   (a) A parabolic mirror reflects a parallel beam of light to a common point, its focus. (b) An elliptical mirror reflects light emitted at one focus onto the other focus. (c) A hyperbolic mirror reflects light from one focus so that it seems to come from the other.

Figure 24(a) shows a parabolic mirror. Such a mirror will focus incoming parallel rays of light or it will send out light in parallel rays if the source of light is at the focus. Figures 24(b) and (c) illustrate the focusing properties of elliptical and hyperbolic mirrors. Reflecting telescopes usually include two mirrors, the larger reflector in such a telescope usually being parabolic and the small one often being hyperbolic. The mirrors are used to reflect the image to the eyepiece. Whispering galleries usually have an elliptical ceiling so that a person standing at one focus can hear a slight noise made at the other focus while someone standing between the two foci can hear nothing. The hyperbola can be used to aid navigation and mapping with radio waves.

There are many other interesting cases of conic sections. The planets move in elliptic orbits with the Sun at a focus; the cable of a suspension bridge hangs approximately in the form of an arc of a parabola; in economics and business, certain types of parabolic and hyperbolic curves are appropriate for representing demand and supply, production and many other relationships.

## 3.1  The parabola

Figure 25 shows the shape of a parabola. The line about which the curve is symmetrical is called the *axis* and the point where the curve cuts the axis is called the *vertex*.

33

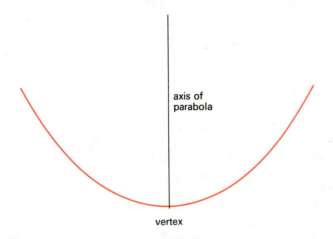

*Figure 25 The general shape of a parabola.*

If the vertex is at the origin and the axis of the parabola lies along the $Y$-axis, as in Figure 26(a), then the parabola will have an equation of the form

$$Y = AX^2$$

where $A > 0$. If the curve opens downward, as in Figure 26(b) then $A < 0$.

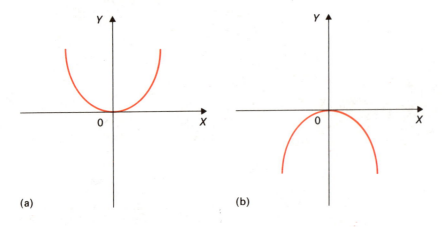

*Figure 26 (a) The parabola $Y = AX^2$ ($A > 0$). (b) The parabola $Y = AX^2$ ($A < 0$).*

If the vertex is at the origin and the axis of the parabola lies along the $X$-axis as in Figure 27(a), then the parabola will have an equation of the form

$$X = AY^2$$

where $A > 0$. If the curve opens to the left instead of the right, as in Figure 27(b), then $A < 0$.

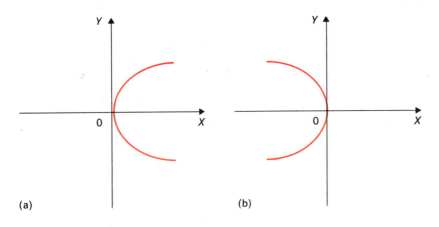

*Figure 27 (a) The parabola $X = AY^2$ ($A > 0$). (b) The parabola $X = AY^2$ ($A < 0$).*

### SAQ 18

On the same axes plot graphs of

(a)   $Y = 2X^2$

(b)   $Y = 4X^2$

for values of $X$ from $-3$ to $+3$.

34

As SAQ 18 has illustrated, the larger the value of $A$ in the equation

$$Y = AX^2$$

the smaller the spread of the parabola. The same is true of parabolas whose equations are of the form $X = AY^2$.

While you were studying the circle in Section 2 you learned that when the centre of the circle shifted from the origin to $(X_1, Y_1)$, $X$ was replaced by $(X - X_1)$ and $Y$ by $(Y - Y_1)$ in the general equation of the circle. You may guess, if the vertex of the parabola is shifted from the origin to $(X_1, Y_1)$, $X$ is again replaced by $(X - X_1)$ and $Y$ by $(Y - Y_1)$, so the equation is

$$Y - Y_1 = A(X - X_1)^2 \qquad \text{(axis parallel to the $Y$-axis)}$$

or

$$X - X_1 = A(Y - Y_1)^2 \qquad \text{(axis parallel to the $X$-axis)}$$

**SAQ 19**

SAQ 19

In terms of $A$, what are the equations of each of the parabolas shown in Figure 28? In which of the parabolas is $A$ positive?

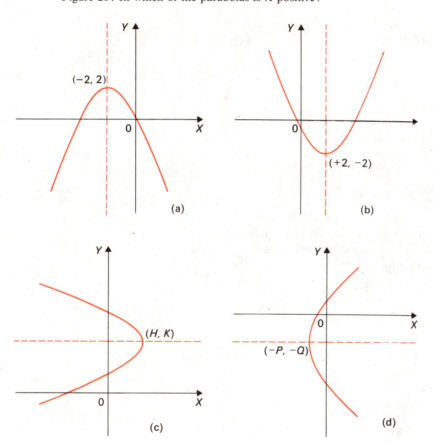

*Figure 28   See SAQ 19.*

### 3.2   Ellipses and hyperbolas

**The ellipse**

If you were to stick two drawing pins into a drawing board, fasten each end of a loose piece of string to the pins, insert a pencil through the string, draw it taut and then draw the shape as you moved the pencil, keeping the string taut, the shape you would obtain is an *ellipse*. You might use this technique to mark out an oval flowerbed.

Figure 29 shows an ellipse centred at the origin. The points $F_1$ and $F_2$ are the foci and correspond to the drawing pins in the contruction I discussed above. Since the string was of a fixed length, it is clear that for any point P on the ellipse the distance $PF_1 + PF_2$ must be constant.

35

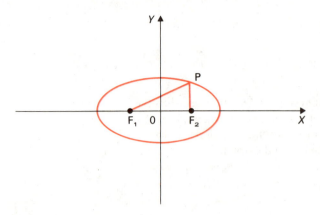

*Figure 29    An ellipse with foci $F_1$ and $F_2$ whose centre is at the origin. Its equation is $X^2/A^2 + Y^2/B^2 = 1, (A \neq B)$.*

The general equation of an ellipse centred on the origin is

$$\frac{X^2}{A^2} + \frac{Y^2}{B^2} = 1 \qquad (A \neq B)$$

If $A > B$ then the ellipse is like the curve shown in Figure 29: if $A < B$ then the ellipse is turned through 90° so that its foci are on the $Y$-axis.

What happens if $A = B$?

The equation is then of the form

$$\frac{X^2}{A^2} + \frac{Y^2}{A^2} = 1$$

which is $X^2 + Y^2 = A^2$ and the curve is a circle.

The fact that the coefficients of $X^2$ and $Y^2$ are *different*, but both positive, is the way of distinguishing an ellipse from a circle. The more nearly $A$ and $B$ are equal, the more nearly circle-like the ellipse becomes; when $A$ and $B$ are very dissimilar, the ellipse is very long and thin.

**SAQ 20**                                                     SAQ 20

(a)   In the ellipse

$$\frac{X^2}{9} + \frac{Y^2}{16} = 1$$

what are $A$ and $B$? Write the equation with $Y^2$ as the subject and hence tabulate values of $Y$ for values of $X$ from $-3$ to $+3$. Plot the ellipse.

(b)   In the ellipse

$$\frac{X^2}{16} + Y^2 = 1$$

what are $A$ and $B$? Tabulate values of $Y$ for values of $X$ from $-4$ to $+4$ and hence plot the ellipse.

**The hyperbola**

The fourth and last of the conic sections is the hyperbola. A hyperbola centred on the origin has an equation of the form

$$\frac{X^2}{A^2} - \frac{Y^2}{B^2} = 1$$

and Figure 30 shows the shape of a hyperbola. It differs from the other conic sections in that it has two distinct and separate branches. (Remember from

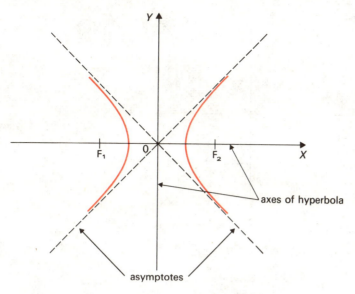

axes of hyperbola

asymptotes

Figure 1 that it is formed by cutting a slice downwards through the two cones placed vertex-to-vertex; there are two branches because such a slice cuts through each of the two cones.) $F_1$ and $F_2$ are the two foci, and a property of the hyperbola is that for any point P on the curve the difference in the lengths $PF_1 - PF_2$ is constant.

The distinguishing feature between the equation of a hyperbola and the equations of the other conic sections is that the coefficients of $X^2$ and $Y^2$ are of a *different* sign: for hyperbolas of the general shape shown in Figure 30, the coefficients of $X^2$ are positive, but those of $Y^2$ are negative.

How do you think a hyperbola of the form

$$\frac{Y^2}{B^2} - \frac{X^2}{A^2} = 1$$

would differ from the one shown in Figure 30?

Its branches would be around the $Y$-axis rather than the $X$-axis.

In Figure 30, I have drawn in two broken lines. You can see that the two branches of the hyperbola seem to nestle in between the pair of lines without touching them. In fact, however far the hyperbola is extended it *never* quite touches these lines. The lines are called the *asymptotes* of the hyperbola.*

asymptote

As you might expect by now, if an ellipse or hyperbola is not centred on the origin, but on the point $(X_1 Y_1)$ then $X$ becomes replaced by $(X - X_1)$ and $Y$ by $(Y - Y_1)$ in the general equations:

For an ellipse

$$\frac{(X - X_1)^2}{A^2} + \frac{(Y - Y_1)^2}{B^2} = 1 \qquad (A \neq B)$$

and for a hyperbola

$$\frac{(X - X_1)^2}{A^2} - \frac{(Y - Y_1)^2}{B^2} = 1$$

**SAQ 21**

**SAQ 21**

Identify each of the following curves as a circle, parabola, ellipse or hyperbola.

(a)  $\dfrac{X^2}{2} - Y^2 = 1$

(b)  $2X^2 + 2Y^2 = 5$

(c)  $X^2 = 3Y$

(d)  $X^2 + 3Y^2 = 4$

(e)  $X^2 = 1 + Y^2$

*This name is not specific to hyperbolas, but can be used for any straight line to which a given curve approaches closely without actually touching or cutting it.*

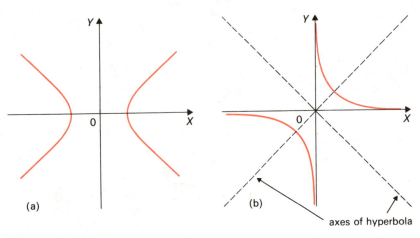

*Figure 31   (a) The hyperbola $X^2 - Y^2 = 1$. (b) The same hyperbola rotated through 45°: its equation is now $XY = 1$.*

A point to notice about all three of the conic sections discussed in this section is that they have all been drawn with their axes parallel to the $X$ or $Y$-axes. Figure 31(a) shows one such hyperbola. Its equation is $X^2 - Y^2 = 1$, which is in the standard form with $A = B = 1$. Figure 31(b) shows the same hyperbola turned so that it fits into the first and third quadrants and its axes are at an angle of 45° to the $X$ and $Y$-axes—its equation is now $XY = 1$.

Such a hyperbola is known as a *rectangular hyperbola*.

What are its asymptotes?

The $X$ and $Y$-axes.

Rotating the hyperbola shown in Figure 31(a) to the position shown in Figure 31(b) has a remarkable effect on its equation. Indeed, this is the first equation of a conic section that I have introduced, that contains an $XY$ term. It is not necessary for you to be able to recognize equations of conic sections whose axes do not lie parallel to the $X$ and $Y$-axes, but notice that an equation which does not *appear* to be that of a conic section may well be the equation of a conic section which is twisted with respect to the $X$ and $Y$-axes.

### 3.3   Summary

For conic sections with axes parallel to the $X$ and $Y$-axes and centred on $(X_1, Y_1)$ the standard equations are:

*Parabola*
$$Y - Y_1 = A(X - X_1)^2$$
If $A > 0$ the curve opens upward and if $A < 0$ the curve opens downward.
$$X - X_1 = A(Y - Y_1)^2$$
If $A > 0$ the curve opens to the right and if $A < 0$ the curve opens to the left.

*Ellipse*
$$\frac{(X - X_1)^2}{A^2} + \frac{(Y - Y_1)^2}{B^2} = 1 \qquad (A \neq B)$$

*Hyperbola*
$$\frac{(X - X_1)^2}{A^2} - \frac{(Y - Y_1)^2}{B^2} = 1$$

# 4  DIRECTED QUANTITIES—VECTORS

## 4.1  Describing position

If a visitor to the Open University were to call me from the gatehouse and ask the way to my office, I would become involved in descriptions like 'turn left, through the car park, then follow the perimeter road to the Science/Technology building' and so on. From my point of view, the details of the description would not matter: any one of many routes would bring the visitor to my office. My visitor, of course, would not only be concerned with directions, but with lengths, especially if he were walking and my suggested route took him the long way around the perimeter road.

Among all the possible routes I might describe, two things would be fixed— the gatehouse would always be the starting point and my office would be the finishing point. So long as attention is restricted to just the initial and final points of the routes, however, it would be fair to describe the routes as equivalent. I might describe all the routes by merely noting the starting and finishing points. It would not do of course just to name the points, because *direction* is important in all routes: the route must make it clear whether the visitor is to go from the gatehouse to my office or from my office to the gatehouse.

Now consider an abstract representation of what I am talking about (Figure 32). The initial position of my visitor is the origin, that of my office is A. Points B, C, D are various landmarks that distinguish between various routes. The stages of the routes are indicated by arrows. Thus the position of each point relative to the last one is represented by an arrow—the length of each arrow is proportional to distance and its direction gives the direction of each point with respect to the starting point of the arrow.

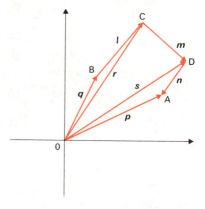

*Figure 32  The distance and direction of one point with respect to another may be represented by an arrow.*

Rather than refer to each point by its co-ordinates, I will use a single simple label which will simultaneously refer to a distance and a direction. For example, the position of my office is given by the length of OA and the direction from O to A. The position of B is specified by the length of OB directed from O to B. These directed lengths essentially refer to *changes in position*, or *displacements*.

I can specify $\overrightarrow{OA}$ as the label for the arrow from O to A, and $\overrightarrow{OB}$ as the label for the arrow from O to B. Both $\overrightarrow{OA}$ and $\overrightarrow{OB}$ then represent directed lengths. They are quantities which cannot be described in terms of size alone: it takes both *direction* and *magnitude* to represent them fully. Such quantities are examples of *vectors*. Quantities that are specified by just their size, for instance, the distance from my office to the gatehouse, as opposed to the route, are called *scalars*.

**vectors**

**scalars**

### Study comment

The notation $\overrightarrow{OA}$ is sometimes used for a vector indicating a change of position from O to A. In this course, however, I shall use a different notation. For example, I will write *p*, *q*, *l*, instead of $\overrightarrow{OA}$, $\overrightarrow{OB}$, $\overrightarrow{BC}$ and so on.

In print, there is a convention that **bold** face type distinguishes a vector quantity— one that has direction as well as magnitude—from scalar quantities, which have only magnitude. For example, *p* indicates the scalar distance to my office, while *p* indicates the change in position, which involves distance and direction. This is an awkward convention when you want to write things by hand, so I suggest that you write the vectors *p* and *q* as $\vec{p}$ and $\vec{q}$, or as $\overrightarrow{OA}$ and $\overrightarrow{OB}$.

Write down vectors representing $\overrightarrow{CD}$ and $\overrightarrow{DA}$ using similar notation to that for $\overrightarrow{OA}$ and $\overrightarrow{OB}$.

You could write $\vec{m}$ and $\vec{n}$ for these vectors (Figure 32).

## 4.2 Adding up vectors

There are several ways in which the route from O to A can be specified. The longest route that is shown involves proceeding from O to B, from B to C, from C to D and finally from D to A. I shall refer to $\overrightarrow{BC}$ by $l$, to $\overrightarrow{CD}$ by $m$, and to $\overrightarrow{DA}$ by $n$. Then the longest route may be represented by

$$q + l + m + n$$

The end result of this is the same as that of the shortest route $\overrightarrow{OA} = p$. Therefore

$$q + l + m + n = p.$$

In words, I can say that $p$ is the *vector sum* of $q$, $l$, $m$ and $n$.

vector sum

As used in this equation, the $+$ and $=$ signs have different meanings to those used when adding or equating numbers. By $q + l + m + n$, I mean the process of combining $q$, $l$, $m$ and $n$ by putting them nose-to-tail as in Figure 32. The equivalence of $q + l + m + n$ to $p$ means that I can close the series of arrows formed by putting $q$, $l$, $m$ and $n$ end-to-end with an arrow drawn from O to A.

### SAQ 22

**SAQ 22**

By considering the various routes from O to A in Figure 32, write down all the alternative combinations of $q$, $l$, $m$, $n$, $r$ and $s$ that result in $p$.

There is one other convention about the use of vectors that I need to introduce. This is that I shall treat vectors as equal when they have the same magnitude and direction. Thus in Figure 33 the vector $\overrightarrow{O'A'}$, which is of the same length as OA and has the same direction, will be described by the same displacement vector, $p$, as $\overrightarrow{OA}$. Similarly, all the other vectors shown will be described by $p$.

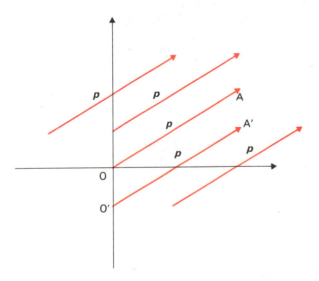

*Figure 33 All the vectors shown are equal since they are of equal size and point in the same direction.*

Used in this way, these vectors relate different starting points to different finishing points; since, for example, O' does not represent the gatehouse, A' cannot be the position of my office. Given just the vector for his route, my visitor would not end up at my office, point A, if he did not start from the point O. However, the point A' bears the same relationship to O' (in distance and direction) as A does to O.

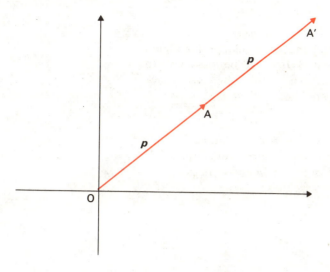

Figure 34  Since $OA = p$ and $AA' = p$, $OA' = 2p$.

The nose-to-tail law for the addition of vectors gives a simple law for the addition of equal vectors. Figure 34 shows the result of putting two vectors, both of which can be represented by $p$, nose-to-tail. The point A' is exactly twice as far away from O as A and is in the same direction. The arrow from O to A' can be represented by $2p$ so that $p + p = 2p$.

## SAQ 23

SAQ 23

Figure 35 shows the plan of roads in a city built according to an idealized rectangular grid pattern with an east–west spacing $a$ and a north–south spacing $b$.

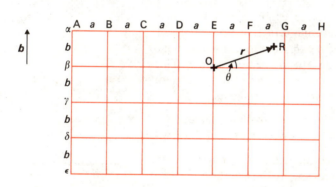

Figure 35  See SAQ 24.

If the north–south roads are labelled A, B, C, D, E, F, G, and H and the east–west roads are labelled $\alpha$ (alpha), $\beta$ (beta), $\gamma$ (gamma), $\delta$ (delta) and $\epsilon$ (epsilon), then each road junction may be labelled. For example, the junction marked O may be labelled $(E, \beta)$. Let the displacement vector from $(A, \alpha)$ to $(B, \alpha)$ be $a$ and the displacement vector from $(A, \beta)$ to $(A, \alpha)$ be $b$. Starting at O, what junctions are reached according to the following vectors?

(a)  $l = a + b$

(c)  $n = 2a - 2b$

(b)  $m = -3b + 2a$

(d)  $p = -a - 3b$.

## SAQ 24

SAQ 24

Consider that the bus routes in the city represented in Figure 35 run along each east–west road and along each north–south road. Every time you want to change direction, you need to change your bus. What is the minimum number of bus changes that are required to reach any road junction in the city? Write down the vector combinations representing the possible bus journeys from O to junction $(F, \delta)$ that involve the minimum number of changes.

## The components of a vector

Consider another point, R, which lies within the rectangular grid of Figure 35 and has a vector *r* relative to the point O. It would be reasonable to suppose that it could be written as the sum of two other vectors or *components* parallel to *a* and *b*. If these are *x* and *y* then

$$r = x + y.$$

The component *x* could be 1.5*a* and the component *y* could be 0.7*b*, for example, leading to the point R. I have drawn *r* in Figure 36 by starting at O and constructing a right-angled triangle of base 1.5*a* and of height 0.7*b*.

vector components

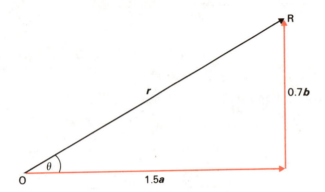

*Figure 36   The vector **r** can be described in terms of two components at right-angles to each other. These components are 1.5a = r cos θ and 0.7b = r sin θ.*

When displayed in this form, it is not difficult to see that if the length of the vector is *r* and the angle that its direction makes with the horizontal is *θ* (theta), the perpendicular sides of the triangle are of length $r \cos \theta$ and $r \sin \theta$ respectively. Therefore, in this instance,

$$r \cos \theta = 1.5a$$
$$r \sin \theta = 0.7b$$

To find *θ*, I can note that division of these expressions gives $\tan \theta = 0.7b/1.5a$, so that *θ* is the angle whose tangent is $0.7b/1.5a$. For the case of a city with an equal spacing of all roads, $a = b$, the angle *θ* is the angle whose tangent is $0.7/1.5 = 0.467$; so $\theta = 25°$ (to the nearest degree). This way of splitting up a vector is known as *resolution* and the act of splitting up a vector is known as resolving the vector into its components.

resolution of a vector

Of course the reverse process is possible also. Given the components of a vector it is possible to combine them to give the vector itself.

Suppose that the horizontal and vertical components of a vector were of magnitude 4 and 3, respectively. These values specify the lengths of the two shorter sides of a right-angled triangle (Figure 37). The longest side, the hypotenuse, then represents the magnitude and direction of the vector resulting from the combination of these components.

By Pythagoras' theorem, the square of this length is equal to $3^2 + 4^2$. This means that the magnitude of the vector is $\sqrt{25} = 5$.

How can I calculate the direction of this vector?

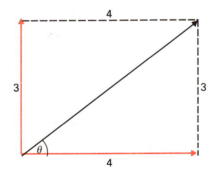

*Figure 37   A vector **r** with 'horizontal' and 'vertical' components 4 and 3 is of magnitude r = √(3² + 4²) = 5. The tangent of the angle the vector makes with its 'horizontal' component is 3/4.*

This is specified by the angle *θ* in Figure 37, the tangent of which is 3/4 or 0.75. By use of trigonometric tables or by use of the slide rule, this angle is 37°, to the nearest degree.

### SAQ 25

SAQ 25

A vector represents a distance of 5 km in a direction 10° south of west. Express it as a sum of components along the directions east and north.

The horizontal component of a vector has a magnitude of 10. The vector itself has magnitude 20. What is the magnitude of the vertical component of the vector and what angle does the vector make with respect to the vertical?

## 4.3 Velocity, acceleration and force

So far I have discussed examples in which vectors were used to describe the position of one point relative to another. A vector representation is, however, appropriate to any quantity whose direction is important and which obeys the nose-to-tail addition law. In this present section, I shall apply similar arguments to some simple problems involving another vector— *velocity*, which describes motion in a particular direction. I shall also introduce the ideas of acceleration and force, both of which are quantities related to directed movement. I do not intend to treat these ideas exhaustively in this section, but would like you to think of it as a vocabulary section that will give you a chance to become familiar with the ideas of acceleration and force before going on to use them in later units.

### Velocity

You may remember that in Unit 2 you had to deal with problems involving the speed of a train. In these problems, you worked out the speed of a train by finding out how long the train took to travel a given distance, You then related the speed, $s$, to the distance, $d$, and the time, $t$, by the equation

$$s = d/t$$

In the context of this problem, you were not interested in the details of the direction in which the train was going: the assumption was that it ran on a track running from Marseilles to Paris and your need was only to know whether the train was heading towards Marseilles or towards Paris.

When you travel by foot or by car, the guiding rails are no longer there and you have to know more about the direction in which you are going. It is when the direction of travel becomes an essential part of your considerations of motion that you need to turn from a description in terms of speed to a description in terms of velocity. Figure 38 shows a potential collision situation: clearly, the precise details of both speed and direction determine whether the vehicles make contact or go past one another.

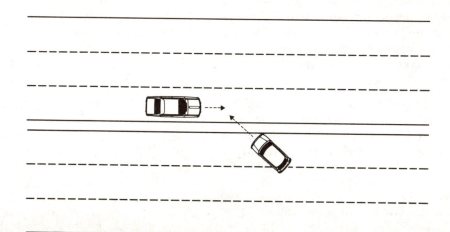

*Figure 38    To determine whether the cars will collide, both the speeds and the directions in which they are travelling must be known.*

43

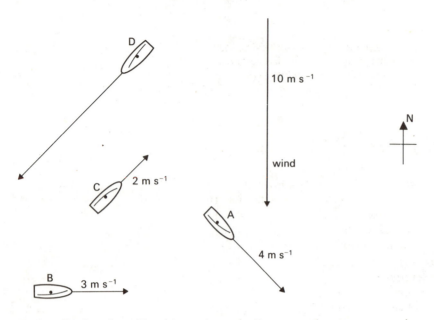

Figure 39 The velocities of the wind and yachts are represented by arrows whose lengths are proportional to the speeds shown.

To handle situations like this mathematically, we need a vector quantity called *velocity*. Graphically, velocity is easy to represent. It is shown as an arrow pointing in the direction of motion with a length proportional to the speed in the same way as you used arrows to denote displacement. Figure 39 shows this kind of representation applied to three yachts A, B and C and also to the wind which drives them. The wind is shown with a velocity of ten metres per second ($m\,s^{-1}$) coming from the north: the yachts have velocities of four metres per second to the south-east, three metres per second to the east and two metres per second to the north-east, respectively.

velocity

### SAQ 27

A fourth yacht, D, is shown in Figure 39. Its velocity is represented by an arrow. What is this velocity?

SAQ 27

### SAQ 28

Represent the collision situation of Figure 38 by a vector diagram. Assume that the vehicle crossing the central reservation has a velocity of 30 metres per second at an angle of 20° to the motorway, and the other vehicle has a velocity of 25 metres per second along its track of the motorway.

SAQ 28

Before attempting to represent a velocity in print, let me just return for a moment to the idea of speed. The equation $s = d/t$ effectively defines speed as a rate of covering distance: if I measure in metres and seconds, then in every second a train moving with speed $s$ metres per second covers a distance $d$ metres; and in every two seconds it covers a distance $2d$ metres. Speed therefore has the character of a rate: it represents something which happens in one unit of time.

To fix the *velocity* of the train I have to work, not in terms of a distance, but in terms of a displacement vector. To perform the equivalent of the speed calculation, I shall assume that the train changes position by an amount that can be described by a displacement vector $\boldsymbol{d}$ in a time $t$. Its velocity is,

$$\boldsymbol{v} = \boldsymbol{d}/t$$

The analogy of velocity with speed works then if I just think of velocity as *a rate of change of position*, where position is to be seen as a vector quantity of the sort described in Section 4.1.

This may sound a formidable sort of definition, but the ideas involved are, in fact, quite simple. The main points to notice are concerned with *dimensions* and with *direction.* In the equation for velocity, **d** is a distance measured in some specified direction, so it has the dimension of length, just as the distance *d* had in the equation for speed. The velocity **v** therefore has the dimensions of length/time, exactly as the speed *s* had in the earlier equation.* It also follows that the direction of the velocity is the same as the direction of the displacement: if the velocity is the rate of change of position, it is also the displacement which takes place in unit time. If they are vectors, they can only be equal if they have the same direction.

As vectors, velocities may be added using the nose-to-tail rule and they can also be resolved into components. The rules for handling velocity vectors are just the same as those you met for handling displacement vectors.

The small flame-cutter of Figure 40 is an instrument for which a description in terms of velocity is appropriate. These instruments have two drives. When just one drive is used the cutter moves in a straight line: when only the other drive is used the cutter moves along a line perpendicular to the first line. Individual control of the two drive systems determines the direction in which the cutter moves.

*Figure 40    A flame cutter is moved simultaneously in two directions at right-angles. These two drives determine the overall motion of the cutter. (British Oxygen Company.)*

Suppose that a machine like this could cut a particular quality of steel sheet at a maximum rate of $0.2\,\text{mm s}^{-1}$ (millimetres per second). How should the controls be set so as to make it follow a track at an angle of 20° to one of the principal drive directions at the maximum cutting rate?

Let me designate the drive directions *x* and *y* and suppose that the required direction of movement is between the $+x$ and the $+y$-directions at an angle of $\theta$ to the $+x$-direction (Figure 41) where $\theta = 20°$ in this instance. If I use **v** to represent the required velocity vector of the cutter, **v** will have a magnitude of $0.2\,\text{mm s}^{-1}$ and a direction set at 20° to the $+x$-direction.

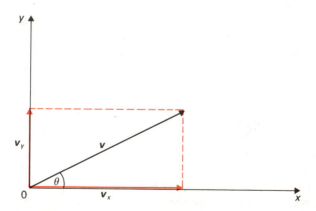

*Figure 41    $\cos\theta = v_x/v$ and $\sin\theta = v_y/v$.*

*Since velocity has dimensions of length/time a common unit used to measure velocity is the metre per second. This can be written as m/s or as $\text{m s}^{-1}$, where the $-1$ is used in a similar way to the negative exponents you met in Unit 3. In this course, $\text{m s}^{-1}$ will be used in preference to m/s.*

What the control settings of the cutter fix are the velocity components in the $x$ and $y$-directions. To find the control settings I need to find the components of $\boldsymbol{v}$ in the $x$ and $y$-directions. These components I shall designate $\boldsymbol{v}_x$ and $\boldsymbol{v}_y$, and their magnitudes are simply

$$v_x = v \cos 20° = 0.2 \cos 20° \text{ mm s}^{-1}$$

and

$$v_y = v \sin 20° = 0.2 \sin 20° \text{ mm s}^{-1}$$

Thus

$$v_x = 0.19 \text{ mm s}^{-1}$$
$$v_y = 0.068 \text{ mm s}^{-1}$$

are the values that must be set on the $x$-movement and $y$-movement speed controls.

### SAQ 29

SAQ 29

Suppose the flame cutter was set to have velocity components $\boldsymbol{v}_x = 0.2 \text{ mm s}^{-1}$ and $\boldsymbol{v}_y = -0.3 \text{ mm s}^{-1}$. What velocity vector would describe its motion? Would this vector be compatible with a maximum cutting speed of $0.5 \text{ mm s}^{-1}$?

### Acceleration

When you drive a car you need constantly to be changing your speed. If your instincts are sporting ones, you probably look for a car which would be described as having rapid acceleration and good brakes: that is, you want a car whose speed can be increased rapidly and then reduced rapidly. If you are unfortunate enough to be involved in a crash, a sideways collision is likely to make your car start to move sideways. It thus accelerates sideways instead of in the usual forward direction. You can appreciate that acceleration concerns changes in speed and is also concerned with direction. Thus the characteristic of acceleration is that it involves a change in velocity.

In every-day life, no attention is paid to the vector aspect of acceleration. A car's performance is quoted as '0–50 mph in ten seconds' or as '0–80 kph in ten seconds'. These figures are interpreted as meaning that, if the car increases its speed by 80 kilometres per hour in ten seconds, it can be expected to increase its speed by 8 kilometres per hour in one second. Its acceleration, which is considered as its rate of changing speed, is then 8 kilometres per hour per second.

This is in unpleasantly mixed units, but if I change the car's specification to '0–22 metres per second in ten seconds' then it changes its speed by $2.2 \text{ m s}^{-1}$ in each second, so that its acceleration can be quoted as $2.2 \text{ m s}^{-1}$ per second, which is written as $2.2 \text{ m s}^{-2}$. As you can see from the units it is measured in, acceleration has the dimensions of length/time². (If you use the negative exponent, this is (length) (time)$^{-2}$, which explains the notation of $\text{m s}^{-2}$ for the units of acceleration.)

It is implicit in the specification of a car's performance that the acceleration takes place in the forward direction and it is therefore considered as a rate of change of speed. Previously, however, I said that acceleration also involves direction and to bring in the idea of direction and so to turn acceleration into a vector, I am going to define *acceleration* as a *rate of change of velocity*. Thus, things which accelerate have different velocity vectors after acceleration to the ones they have before. The difference between the above definition of acceleration and the colloquial use of the word is that it allows us to refer to all rates of change of velocity, even those which involve no change in speed, as accelerations. During a rebound or when a vehicle

*acceleration*

follows a curved path, there may be no change in speed, but the direction, and hence the velocity vector, is changing. Thus, by definition, acceleration is taking place.

The type of motorway collision portrayed in Figure 38 can be prevented by inserting a crash barrier into the central reservation. What description of the barrier's influence can be made using velocity vectors? As discussed in Road Safety (one of the essays from *Modelling Themes*) the barrier should bend under impact so that the vehicle is deflected at as small an angle as possible to the line of the barrier. Figure 42 shows a collision with a crash barrier which results in a change in the velocity (vector) of the crashing vehicle.

Suppose that the speed of the vehicle before and after impact is $40 \, \text{m s}^{-1}$. This means that the speed is unaffected by the impact.

Has the vehicle's velocity changed?

Because the velocity after impact has a different direction to the velocity before impact, the velocity has changed, even though the speed is the same. The way in which the velocity has changed is best seen by resolving it into components. A convenient reference direction is the line of the crash barrier before impact.

If the initial velocity was $40 \, \text{m s}^{-1}$ at an angle of $\theta$ to the crash barrier, the initial components of velocity are $40 \cos \theta \, \text{m s}^{-1}$ from left-to-right along the crash barrier and $40 \sin \theta \, \text{m s}^{-1}$ at right-angles to the barrier and *towards* it.

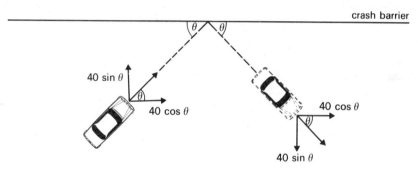

*Figure 42  The result of a car crashing into a crash barrier can be deduced by resolving its motion into perpendicular components.*

Since the vehicle is reflected at the same speed and angle, the final components of velocity are $40 \cos \theta \, \text{m s}^{-1}$, again from left-to-right along the crash barrier and $40 \sin \theta \, \text{m s}^{-1}$ at right-angles to the barrier, but this time *away* from it.

In this case the component of velocity along the crash barrier is unchanged and it is only the component at right-angles to the barrier that changes. The final perpendicular component is equal in magnitude, but opposite in direction, to the initial perpendicular component. The vector that is equal in magnitude, but opposite in direction to vector *a* is represented by $-a$, so the change from *a* to $-a$ is given by

$$a - (-a) = 2a$$

In other words, the change is a new vector of magnitude 2*a*. For the crashing vehicle I can write

final velocity − initial velocity = 2*a*.

Therefore, the magnitude of the change in velocity is

$$40 \sin \theta - (-40 \sin \theta) \, \text{m s}^{-1} = 80 \sin \theta \, \text{m s}^{-1}.$$

The direction of this velocity change is perpendicular to, and away from, the crash barrier.

Suppose that the car was in contact with the barrier for two seconds and that the angle was 20°. Then the change in velocity is $80 \sin 20° = 27\,\mathrm{m\,s^{-1}}$ directed away from the barrier and takes place over a period of two seconds. If the change takes place uniformly over this period, the rate of change of velocity—that is, the change in velocity per second—is $27/2\,\mathrm{m\,s^{-2}} = 13.5\,\mathrm{m\,s^{-2}}$. The acceleration is thus $13.5\,\mathrm{m\,s^{-2}}$ at right-angles to the barrier and directed away from it.

What I have done in this example is just to apply the rate of change idea to a velocity in order to deduce an acceleration. I can write an equation for this. Let something have an initial velocity $v_i$ and, at a time $t$ later, a final velocity $v_f$. If it accelerated uniformly during the time $t$ its acceleration will be

$$a = \frac{v_f - v_i}{t}$$

### SAQ 30

SAQ 30

A car moving in a straight line changes velocity from $12\,\mathrm{m\,s^{-1}}$ to $24\,\mathrm{m\,s^{-1}}$ in seven seconds. What is its average acceleration over the period?

What would be the acceleration if the car slowed from $24\,\mathrm{m\,s^{-1}}$ to $12\,\mathrm{m\,s^{-1}}$ in the same period?

Notice that the acceleration in the second part of SAQ 30 turns out to be negative and slowing down is therefore often called negative acceleration. It can also be called deceleration or retardation.

It does not take an impact against a crash barrier to make a vehicle accelerate in a direction other than along its line of motion. Much more ordinary circumstances will do that. Suppose you were to drive a car around a corner at constant speed. Would you say that the car was accelerating? Conventionally, you would not, but according to my definition of acceleration as a rate of change of velocity, you would have to. A car which changes its direction of motion changes its velocity and velocity change means acceleration.

If the car were to follow a curve at a fixed speed of $20\,\mathrm{m\,s^{-1}}$ and, in doing so, turned through an angle of 30° in a period of three seconds, what acceleration would it experience?

Figure 43(a) represents the situation. The initial velocity vector of the car is $v_i$ and the final vector is $v_f$. These vectors are shown with the same magnitude and are drawn from the same origin at an angle of 30° to one another (Figure 43(b)). The vector $v$ represents the vector difference between $v_f$ and $v_i$, since the vector addition rule gives

$$v_f = v_i + v$$

Since the magnitudes of $v_f$ and $v_i$ are equal, the triangle representing the vectors is isosceles and the magnitude of $v$ must be $2v_i \sin 15° = 2 \times 20 \times 0.258\,\mathrm{m\,s^{-1}} = 10.3\,\mathrm{m\,s^{-1}}$. Its direction, from the symmetry of the diagram, must be perpendicular to the line bisecting the angle between the vectors $v_i$ and $v_f$. The average acceleration which has taken place during the three seconds of the turn is thus $10.3/3\,\mathrm{m\,s^{-2}} = 3.43\,\mathrm{m\,s^{-2}}$ in the direction of the vector $v$.

From this example it should be clear that the acceleration which takes place during motion in a circle is in some direction across the line of motion. It is an acceleration that need not be small in magnitude. The precise magnitude

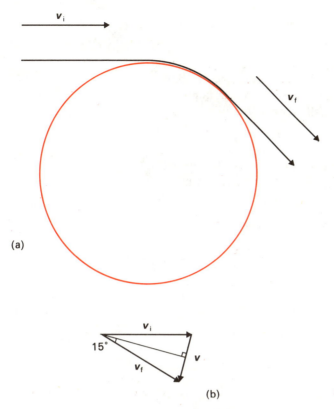

(a)

(b)

Figure 43 (a) Representative of a car travelling around a curve at a steady speed. (b) In the triangle $v_i = v_f$ and $\frac{1}{2}v/v_i = \sin 15°$, where $v$ is the vector difference.

and direction of the acceleration that takes place as a body moves in a circular path is a matter that I shall take up again in detail in Unit 9, which is concerned mainly with motion in a circle.

### Force

If you wanted to uproot an old tree stump from your garden, you would probably begin by loosening the soil around its roots. When it seemed reasonably free you would push it. If that did not work, you might get a friend and he would push too. Perhaps you would put a rope around the stump and one of you would pull the rope while the other pushed the tree, or you might wrap the rope around the stump and each pull an end. In all cases you would be in no doubt of what you were doing: you would be applying *force* to the stump in order to make it move.

force

When you work with someone else on the tree stump, you both push together on the same side of the tree. However, you would not push in precisely the same direction. Because you each need a bit of room, you push against the tree in directions which are somewhat inclined. If you use a rope (Figure 44) you still do not pull in quite the same direction. In either case,

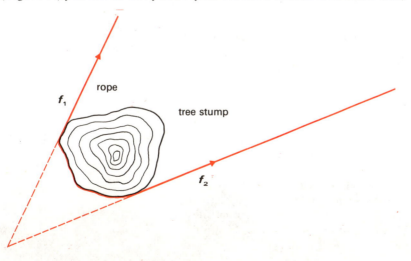

Figure 44 A tree stump being uprooted by pulling on two ends of a rope with forces $f_1$ and $f_2$.

49

however, you will be in no doubt that your combined effect on the tree is greater than what you could do individually.

Given the preceding sections on vectors, it will come as no surprise to you that I want to describe the combination laws for force using vectors. If two men pull on a tree as in Figure 44, I shall describe each man as applying a force to the tree through his rope and I shall combine the two forces by vector addition.

Let me assume that one man pulls with a force $f_1$ and the other with a force $f_2$. The forces are vectors, with lines of action that go through the ropes. To combine the effect of the forces, that is, to find the single force which has the same effect, I shall suppose that I can add them by the usual law of vector addition. Thus (Figure 45) I draw the vector for $f_2$ and add $f_1$ to it to give a combined vector $f$ which I take to be the equivalent single force, or *resultant*, of the other two.

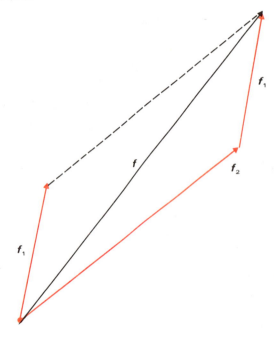

SAQ 31

*Figure 45    The two forces $f_1$ and $f_2$ can be described in terms of a single equivalent force, called the resultant force.*

### SAQ 31

Two men pull, each with a force of magnitude $f$, on a tree. They use ropes which are inclined at an angle of 45°. Write down the resultant force that they produce in terms of the magnitude $f$.

I have deliberately not given any value to the forces because my current interest in force is the fact that it turns out to be yet another quantity of a vector nature. Forces are combined by the laws of vector addition just as are velocities and accelerations.

Just what force is and the units it is measured in are matters that I shall deal with in detail in Unit 6. For the moment, let me just note that the unit of force is the *newton*. One newton is a force just about sufficient to break a piece of fine sewing thread. It is also roughly the force you would need to exert with your hand in order to hold up a small object like an apple.

The symbol for newton(s) is N.

SAQ 32

### SAQ 32

One man pushes a trolley up a short ramp which is at an angle of 30° to the horizontal. He exerts a force of 500 newtons directed up the ramp and he is aided by another man at the top of the ramp who pulls the other end with a rope in a horizontal direction with a force of 400 newtons. What is the resultant force of their efforts?

**Study comment**

The use of trigonometric tables was dealt with in Unit 4, Section 6.3 and on Side 1 of Disc 3.

Side 2 of Disc 3, *Resolving Vectors*, describes the process of adding and splitting up vectors and you should listen to it now.

## 4.4  Summary

1  Vectors are quantities that have both *magnitude* and *direction*. Included among such quantities are displacement, velocity, acceleration and force.

2  A vector can be represented graphically by an arrow whose length is proportional to the magnitude of the vector and whose direction denotes the direction of the vector. In print, a vector is shown in **bold** type—**q**, **p**. The magnitude of a vector is shown in italic type—*q*, *p*.

3  Vectors may be added by a rule in which they are laid end-to-end, the total of a series of such vectors being the single vector which would connect the starting and end points of the vectors being added.

4  A vector may be considered as a sum of two other vectors which are at right-angles to each other, called components. If a vector **p** is at an angle $\theta$ to some specified direction, its components in that direction and in the direction perpendicular to it will be of magnitude $p\cos\theta$ and $p\sin\theta$, respectively.

5  A vector **v** which has components of magnitude $v_x$ along a direction designated $x$, and $v_y$ along a perpendicular direction designated $y$, has a magnitude given by

$$v^2 = v_x^2 + v_y^2$$

and a direction making an angle $\theta$ with the $x$-direction such that $\tan\theta = v_y/v_x$.

6  A velocity is defined as a rate of change of position. It is a vector quantity with the dimensions of length/time and has a direction which is the same as that in which position changes.

7  Where a movement takes place in a straight line, the magnitude of the velocity of a body is its speed.

8  A displacement **d** which takes place uniformly over a time $t$ has a velocity **v** given by $\mathbf{v} = \mathbf{d}/t$.

9  An acceleration is a rate of change of velocity and has the dimensions of length/time$^2$.

10  A body whose velocity changes uniformly from $\mathbf{v}_i$ to $\mathbf{v}_f$ over a time $t$ experiences an acceleration given by $\mathbf{a} = (\mathbf{v}_f - \mathbf{v}_i)/t$.

11  In linear motion, a positive acceleration is one having the same direction as the velocity vector: a negative acceleration has the reverse direction.

12  A body may experience acceleration without changing its speed. This can happen when a body rebounds or when it follows a curved path. In such cases the acceleration is not parallel to the velocity.

13  Forces are vector quantities and may be combined by the usual rules of vector addition. The unit of force is the newton, symbol N.

## SUMMARY OF THE UNIT

In a *rectangular Cartesian co-ordinate system* the axes are at right-angles to each other. It is possible to represent the position of a point in three dimensions by using *three axes and three co-ordinates*. Ordnance Survey maps indicate position by reference to a grid and height by means of *contour lines*.

Section 1.1

It is possible to check whether a triangle whose vertices have known Cartesian co-ordinates is right-angled by using *Pythagoras' theorem*. The *length of each of the sides* can be calculated using the formula

Section 1.2

$$\text{length of side} = \sqrt{[(X_2 - X_1)^2 + (Y_2 - Y_1)^2]}$$

where $(X_1, Y_1)$ and $(X_2, Y_2)$ are the co-ordinates of the points at each end of the side. The *gradient* of a line between $(X_1, Y_1)$ and $(X_2, Y_2)$ can be found from the formula

$$M = \frac{Y_2 - Y_1}{X_2 - X_1}$$

A line which slopes down from left to right will have a *negative gradient*. Lines which are *parallel* have *equal gradients*. Lines which are *perpendicular* to each other are such that the *product of their gradients is equal to* $-1$. It is possible to check if three points P, Q and R are *collinear* by calculating the *gradients* of PQ and QR (or PQ and PR, or QR and PR) and checking they are the same.

The *general equation of a straight line* is

Section 1.3

$$Y = MX + C$$

where $M$ is the *gradient* and $C$ is the *intercept* of the line with the $Y$-axis. The equation of a line with gradient $M$ through a fixed point with co-ordinates $(X_1, Y_1)$ is

$$\frac{Y - Y_1}{X - X_1} = M$$

The equation of a line through the fixed points $(X_1, Y_1)$ and $(X_2, Y_2)$ is

$$\frac{Y - Y_1}{X - X_1} = \frac{Y_2 - Y_1}{X_2 - X_1}$$

A line can be specified *either* by a knowledge of its gradient and the co-ordinates of one point on it, *or* by a knowledge of the co-ordinates of two points on the line.

Von Thunen's agricultural location model (introduced in Unit 2) implies a *circular pattern* of crops with the market at the centre of the circles.

Section 2.1

*The equation for a circle* of centre $(X_1, Y_1)$ and radius $A$, is

Section 2.2

$$(X - X_1)^2 + (Y - Y_1)^2 = A^2$$

which can also be written

$$X^2 + Y^2 - 2X_1 X - 2Y_1 Y = A^2 - X_1^2 - Y_1^2$$

Given the equation for a circle in the above form it is possible to find the centre and radius of the circle by *equating the coefficients* of $X$ and $Y$ and by *equating the constant terms*.

When the quantity $Y$ is *proportional to* the quantity $X$: if $X$ doubles, $Y$ will also double; if $X$ is reduced by a factor of, say, ten $Y$ will also be reduced by a factor of ten; and so on. This relationship can be written as

Section 2.3

$$Y \propto X$$

where $\propto$ is the *proportionality sign*. By introducing a *constant of proportionality*, $K$, this relationship can be written as an equality

$$Y = KX$$

52

Similarly, if $Y$ is *inversely proportional* to $X$ then

$$Y \propto \frac{1}{X}$$

or

$$Y = \frac{K'}{X}$$

where $K'$ is again a constant of proportionality. In a shopping model, the *number of trips made*, $T$, to a shopping centre is supposed to be proportional to the *floor space*, $F \, \text{m}^2$, and the *amount of money*, $£E$, that the occupants of a residential zone will spend and is inversely proportional to the square of the distance, $D \, \text{m}$, from the zone to the centre:

$$T \propto \frac{EF}{D^2}$$

or

$$T = \frac{KEF}{D^2}$$

where $K$ is a constant of proportionality. Using this model it is possible to determine the boundary between areas where shoppers are more likely to go to one shopping centre than another. Such boundaries may be straight lines or circles.

The circle is one example of a *conic section*. Other conic sections are the *parabola*, the *ellipse* and the *hyperbola*. **Section 3**

When a *parabola* has its axis *parallel to the Y-axis* and its vertex at $(X_1, Y_1)$, its equation is **Section 3.1**

$$Y - Y_1 = A(X - X_1)^2$$

If $A > 0$, the curve opens upwards, if $A < 0$, the curve opens downwards. When a *parabola* has its axis *parallel to the X-axis* and its vertex is at $(X_1, Y_1)$, its equation is

$$X - X_1 = A(Y - Y_1)^2$$

If $A > 0$, the curve opens to the right, if $A < 0$, the curve opens to the left. In both cases, the *larger A the smaller the spread of the parabola*.

The general equation of an *ellipse* with its axes parallel to the $X$ and $Y$-axes and its centre at $(X_1, Y_1)$ is **Section 3.2**

$$\frac{(X - X_1)^2}{A^2} + \frac{(Y - Y_1)^2}{B^2} = 1 \qquad (A \neq B)$$

and that of a *hyperbola* with its axes parallel to the $X$ and $Y$-axes and its centre at $(X_1, Y_1)$ is

$$\frac{(X - X_1)^2}{A^2} - \frac{(Y - Y_1)^2}{B^2} = 1$$

A *vector* represents both *magnitude* and *direction*. Vectors are written in **bold** type to distinguish them from *scalar* quantities. *Vectors* can be *added*. For instance, Figure 32 shows that $\boldsymbol{q} + \boldsymbol{l} + \boldsymbol{m} + \boldsymbol{n} = \boldsymbol{p}$. **Section 4.1**

**Section 4.2**

If OA is represented by $\boldsymbol{a}$, then $\boldsymbol{a}$ can also refer to any other vector equal in length to OA and parallel to OA. The vector $-\boldsymbol{a}$ is equal in length to $\boldsymbol{a}$, but points in exactly the opposite direction. The vector $2\boldsymbol{a}$ is twice as long as $\boldsymbol{a}$ and points in the same direction.

A vector can be split into *components*. This process is called *resolving a vector*. If a vector $\boldsymbol{a}$ makes an angle $\theta$ with a direction of interest, its component along that direction has magnitude of $a \cos \theta$. The component at right-angles to that direction has magnitude of $a \sin \theta$. It is also possible to *combine two vectors*. If two vectors $\boldsymbol{a}$ and $\boldsymbol{b}$ are at right-angles to each other then their *resultant* has magnitude $\sqrt{(a^2 + b^2)}$ and makes an angle $\theta$ with the line of $\boldsymbol{a}$ so that $\tan \theta = b/a$.

*Velocity, acceleration* and *force* are vectors. Velocity describes speed in a particular direction and can be described as *rate of change of position*. It is possible for the velocity of an object to change even though its speed is constant. The unit for velocity, the metre per second, can also be written as $m s^{-1}$. Acceleration is the *rate of change of velocity*. The unit for acceleration is the metre per second per second, which is written as $m s^{-2}$. The unit for force is the Newton(N). Since force, acceleration and velocity are all vectors, they can each be resolved or combined in the same way as displacement vectors.

# ANSWERS TO SELF-ASSESSMENT QUESTIONS

## SAQ 1

(a)  Draw in the point at $(0, 2)$ on Figure 7 and call it the point E. For the triangle CEB

$$BE = 2 - 0 = 2$$

and

$$CE = 4 - 2 = 2$$

Since CEB is a right-angled triangle, I can apply Pythagoras' theorem and thus:

$$(BC)^2 = (BE)^2 + (CE)^2$$
$$(BC)^2 = 2^2 + 2^2$$
$$= 8$$

If I now take the point F $(0, 1)$ and apply Pythagoras' theorem to the right-angled triangle CFA:

$$(AC)^2 = (AF)^2 + (CF)^2$$
$$= (1 - 0)^2 + (4 - 1)^2$$
$$= 1 + 9$$
$$= 10$$

(b)  For the triangle ABC

$$(AB)^2 + (BC)^2 = (AC)^2$$

since $2 + 8 = 10$. Thus Pythagoras' theorem does hold for triangle ABC.

## SAQ 2

To answer this question it is necessary to use Pythagoras' theorem.

$$(PQ)^2 = [0 - (-3)]^2 + [2 - (-1)]^2$$
$$= 3^2 + 3^2$$
$$= 18$$
$$(PR)^2 = (0 - 3)^2 + [2 - (-1)]^2$$
$$= (-3)^2 + 3^2$$
$$= 18$$
$$(QR)^2 = [(-3) - 3]^2 + [-1 - (-1)]^2$$
$$= (-6)^2 + 0$$
$$= 36$$

Therefore

$$(QR)^2 = (PR)^2 + (PQ)^2$$

and so PQR is a right-angled triangle, with $Q\hat{P}R = 90°$.

## SAQ 3

Starting with the co-ordinates of R

$$\text{gradient of RS} = \frac{2 - 1}{3 - 0} = \frac{1}{3}$$

Starting with the co-ordinates of S

$$\text{gradient of ST} = \frac{1 - 4}{0 - (-1)} = -3$$

Note that the product of the gradients of RS and ST is $(1/3) \times (-3) = -1$.

$$(RS)^2 = (3 - 0)^2 + (2 - 1)^2 = 10$$
$$(ST)^2 = [0 - (-1)]^2 + (1 - 4)^2 = 10$$
$$(RT)^2 = [3 - (-1)]^2 + (2 - 4)^2 = 20$$

Therefore $(RT)^2 = (RS)^2 + (ST)^2$ and RST must be a right-angled triangle. Note which lines are at right angles.

## SAQ 4

Suppose the co-ordinates of the new point A' are $(K, K)$ and the co-ordinates of the new point B' are $(2K, 2K)$. The gradient of the line through A' and B' is

$$M' = \frac{K - 2K}{K - 2K} = \frac{-K}{-K} = 1$$

which is the same as for the line AB. So the line through the two new points will be parallel to AB.

## SAQ 5

I shall check the gradients of the lines PQ and QR. If they are the same, then since both lines pass through a common point, Q, PQR must lie on a straight line.

$$\text{Gradient of PQ} = \frac{-4}{+2} = -2$$

$$\text{Gradient of QR} = \frac{-2}{+1} = -2$$

So the collinearity is demonstrated.

As a check you could work out the gradient of PR. This gradient is

$$\frac{-3 - 3}{-1 + 4} = \frac{-6}{+3} = -2$$

## SAQ 6

The equation $P = N + AY$ is that of a straight line with gradient $A$. If $P$ and $Y$ are variables than a graph of $P = N + AY$ crosses the $P$-axis at $P = N$; in other words, when $Y = 0$, $P = N$. So $N$ is the intercept with the $P$-axis. The line will cross the $Y$-axis when $P = 0$, that is, when $Y = -N/A$.

## SAQ 7

The equation of the line is determined by the introduction of some point $(X, Y)$. Then the gradient is given by

$$\frac{Y - W}{X - 0} = M$$

To get this equation in the standard form first multiply both sides by $(X - 0)$

$$Y - W = MX$$

Then add $W$ to both sides

$$Y = MX + W$$

So $Y = MX + W$ is the equation of the line in standard form.

## SAQ 8

Consider a point $(X, Y)$ on the line. Then

$$\text{the gradient } M = \frac{Y - (-5)}{X - (-2)}$$
$$= \frac{Y + 5}{X + 2}$$

If $M = -1$, then

$$\frac{Y + 5}{X + 2} = -1$$

Multiplying both sides by $(X + 2)$

$$Y + 5 = -X - 2$$

or

$$Y = -X - 2 - 5$$
$$= -X - 7$$

Thus the standard form of the equation is $Y = -X - 7$.

## SAQ 9

The first step is to calculate the gradient of the line from the given points

$$M = \frac{-3 - 4}{4 - 2} = \frac{-7}{2}$$

Since the gradient is constant for any particular straight line, any other point $(X, Y)$ on the same line should give the same gradient when associated with one of the given points. Therefore

$$\frac{Y - 4}{X - 2} = \frac{-7}{2}$$

To write this in standard form, multiply through by $2(X - 2)$. Then

$$2Y - 8 = -7X + 14$$

and

$$Y = \frac{-7}{2}X + 11$$

and this is the standard form.

## SAQ 10

The gradient of PQ is

$$\frac{-1 - 2}{-3 - 0} = 1$$

Since PR is at right-angles to PQ (PQR is a right-angled triangle with QR as its hypotenuse), the product of the gradient of PR and the gradient of PQ must be $-1$. This means that the gradient of PR must be $-1$. PR must also pass through P $(0, 2)$. Hence

$$-1 = \frac{Y - 2}{X - 0}$$

By rearranging, the equation can be written in the standard form

$$Y = -X + 2$$

## SAQ 11

(a) The equation is that of a circle with $A^2 = 16$ or $A$, the radius, equal to 4.

(b) Divide through the equation by 9

$$X^2 + Y^2 = \frac{1}{9}$$

This is the equation of a circle with $A^2 = 1/9$, or $A$, the radius, equal to 1/3.

## SAQ 12

Rearranging the equation given in the question and taking the general equation of the circle gives

$$X^2 + Y^2 - 8X - 6Y = 0$$

and

$$X^2 + Y^2 - 2XX_1 - 2YY_1 = A^2 - X_1^2 - Y_1^2$$

Comparison with the given equation suggest that

$$-2X_1 = -8$$

so

$$X_1 = 4$$

and

$$-2Y_1 = -6$$

so

$$Y_1 = 3$$

This means that the centre of the circle is $(4, 3)$. Finally

$$A^2 - X_1^2 - Y_1^2 = A^2 - 16 - 9$$
$$= A^2 - 25$$
$$= 0$$

Therefore

$$A = 5.$$

Thus the radius of the circle is 5 and its centre is the point $(4, 3)$.

## SAQ 13

The equation of a circle, with centre $(X_1, Y_1)$ and radius $A$ is

$$(X - X_1)^2 + (Y - Y_1)^2 = A^2$$

In this case, $A = 5$ so

$$(X - X_1)^2 + (Y - Y_1)^2 = 25$$

This is the equation of all circles with radius 5.

## SAQ 14

(a) The number of kilometres distance, $\bar{D}$, from the town centre at which the highest rents for the two crops become equal (to $\bar{R}$, say) is given by

$$\bar{R} = 20\,000 \times 0.03 - 20\,000 \times 0.001\bar{D}$$
$$= 10\,000 \times 0.05 - 10\,000 \times 0.001\bar{D}$$

Therefore

$$\bar{D} = \frac{20\,000 \times 0.03 - 10\,000 \times 0.05}{(20\,000 - 10\,000)0.001}$$
$$= \frac{600 - 500}{10}$$
$$= 10$$

(b) For crop A

$$R_{max} = 600 - 20D$$

and for crop B

$$R_{max} = 500 - 10D$$

where $D$ is the number of kilometres from the town centre.

For $D = 5$, say, crop B will be more profitable than crop A. So crop B will tend to be grown up to a distance of 10 km from the town centre and, therefore, can be expected to be grown at 7 km from the centre.

(c) If you imagine a co-ordinate grid based on kilometre squares, with axes passing through the town centre and the $Y$-axis running SN and the $X$-axis running WE, then the new position of the market can be taken as the point Q $(X_1 Y_1)$ and $X_1 = 5$ and $Y_1 = 10$.

The circular boundary is defined by $A (= \bar{D}) = 10$

So substituting in the general equation for a circle

$$(X - 5)^2 + (Y - 10)^2 = 10^2$$

Therefore

$$X^2 - 10X + Y^2 - 20Y + 25 = 0$$

is the required equation

## SAQ 15

If the shopping centres are equally attractive then for any point P on the boundary of influence

$$\frac{KEF}{D_1^2} = \frac{KEF}{D_2^2}$$

or

$$D_1 = D_2$$

In other words, any point on the boundary is equidistant from the two centres. The locus of P therefore is a straight line running 'vertically' through the mid-point between the two centres, that is, 2 km to the east of the local centre, the point $(0, 2)$.

The equation of this line must be $X = 2$.

## SAQ 16

The change in the ratio of attractiveness means that

$$\frac{1}{D_1^2} = \frac{1}{3D_2^2}$$

for any point on the boundary of influence. However, it is not necessary to work any further to see that whatever the ratio of attractiveness the zone of influence will always be a circle. Altering the ratio of attractiveness merely alters the coefficients of $X^2$ and $Y^2$ in the equation, but does not make them unequal.

The only factor that would alter this would be if the $X$ and $Y$-co-ordinates were given different weightings. That is, if it were more difficult to travel north–south than east–west, then the travel cost would not be calculated simply according to distance (of the form $\sqrt{(X^2 + Y^2)}$), but according to a 'distorted' distance (say $\sqrt{(X^2 + 2Y^2)}$). This will give unequal coefficients of $X^2$ and $Y^2$, and hence a boundary of influence which is not circular.

## SAQ 17

The effect of shifting the local centre to the point $(0, 4)$ will be to shift the centre of the circle of influence.

The new distances would be

$$D_2^2 = X^2 + (Y - 4)^2$$
$$D_1^2 = (4 - X)^2 + Y^2$$

The relationship $2D_2^2 = D_1^2$ would then be

$$2(X^2 + Y^2 - 8Y + 16) = 16 - 8X + X^2 + Y^2$$

or

$$X^2 + Y^2 + 8X - 16Y + 16 = 0$$

## SAQ 18

The curves are plotted in Figure 46.

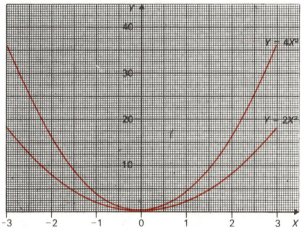

*Figure 46  Graphs of $Y = 4X^2$ and $Y = 2X^2$ are parabolas. As the coefficient of $X^2$ gets larger, the spread of the parabola gets smaller.*

## SAQ 19

(a)  The equation will be of the form

$$Y - Y_1 = A(X - X_1)^2$$

$X_1$ and $Y_1$ are the co-ordinates of the vertex, but the point $(-2, 2)$ is the vertex, so these values can be substituted for $X_1$ and $Y_1$.

$Y_1 = 2$ and $X_1 = -2$, so the equation is

$$Y - 2 = A(X + 2)^2$$

(b)  The equation will again be of the form

$$Y - Y_1 = A(X - X_1)^2$$

$Y_1 = -2$ and $X_1 = 2$ are the co-ordinates of the vertex, so the equation is

$$Y + 2 = A(X - 2)^2$$

(c)  The equation this time will be of the form

$$X - X_1 = A(Y - Y_1)^2$$

and $X_1 = H$, $Y_1 = K$, so the equation is

$$X - H = A(Y - K)^2$$

(d)  The equation will be of the form

$$X - X_1 = A(Y - Y_1)^2$$

and $X_1 = -P$, $Y_1 = -Q$, so the equation is

$$X + P = A(Y + Q)^2$$

$A$ will be positive in (b) and (d).

## SAQ 20

(a)  $A = 3$ and $B = 4$.

The equation can be written

$$Y^2 = 16\left(1 - \frac{X^2}{9}\right)$$

| $X$ | $Y$ |
|---|---|
| $-3$ | 0 |
| $-2$ | $\pm 2.98$ |
| $-1$ | $\pm 3.77$ |
| 0 | 4 |
| 1 | $\pm 3.77$ |
| 2 | $\pm 2.98$ |
| 3 | 0 |

The ellipse is plotted in Figure 47(a).

(b)  $A = 4$ and $B = 1$

| $X$ | $Y$ |
|---|---|
| $-4$ | 0 |
| $-3$ | $\pm 0.66$ |
| $-2$ | $\pm 0.87$ |
| $-1$ | $\pm 0.97$ |
| 0 | $\pm 1$ |
| 1 | $\pm 0.97$ |
| 2 | $\pm 0.87$ |
| 3 | $\pm 0.66$ |
| 4 | 0 |

The ellipse is plotted in Figure 47(b).

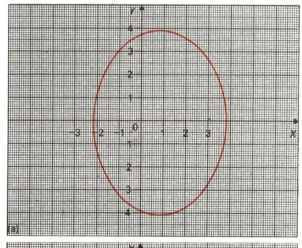

*Figure 47  (a) Graph of $X^2/9 + Y^2/16 = 1$. (b) Graph of $X^2/16 + Y^2 = 1$.*

57

## SAQ 21

(a) The fact that the coefficient of $X^2$ is positive and that of $Y^2$ is negative shows this to be the equation of a hyperbola.

(b) The coefficients of $X^2$ and $Y^2$ are equal and $A^2 > 0$, so this is the equation of a circle.

(c) This is the equation of a parabola.

(d) This equation can be rewritten as

$$\frac{X^2}{4} + \frac{Y^2}{(4/3)} = 1$$

which identifies it as an ellipse.

(e) This equation can be rewritten as

$$X^2 - Y^2 = 1$$

which identifies it as a hyperbola.

## SAQ 22

$q + l + m + n$

$r + m + n$

$s + n$

## SAQ 23

(a) $(F, \alpha)$    (c) $(G, \delta)$

(b) $(G, \epsilon)$    (d) $(D, \epsilon)$

## SAQ 24

Since any junction can be reached by going along an east–west road to the intersection with a north–south road that contains the desired juction, only one bus change is required to reach any point in the city.

The vector combinations for the required bus journey are:

$a - 2b$

$-2b + a$.

## SAQ 25

The direction of the vector is shown in Figure 48. The component in the direction *west* is $5 \cos 10°$ km $= 4.92$ km.

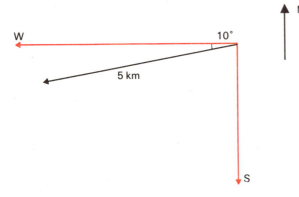

*Figure 48    See the answer to SAQ 25.*

This corresponds to $-4.92$ km in the direction *east*.

The component in the direction *south* is $5 \sin 10°$ km $= 0.868$ km.

This corresponds to $-0.868$ km in the direction *north*.

## SAQ 26

Let the vector make an angle $\theta$ with the vertical. Its horizontal component is then $20 \sin \theta$ and this is known to be 10 (Figure 49). So $\sin \theta = 10/20 = 0.5$ and therefore $\theta = 30°$. The vertical component of the vector is $20 \cos \theta = 20 \cos 30° = 17.3$.

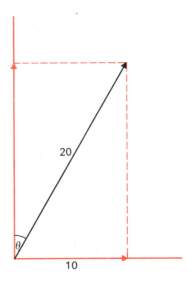

*Figure 49    See the answer to SAQ 26.*

## SAQ 27

The lengths of the arrows represent the magnitude of the vectors and the vector for D is 4/5 the length of the wind vector. Therefore the magnitude of the velocity of D is $4 \times 10/5$ m s$^{-1}$ $= 8$ m s$^{-1}$. Its direction is south–west. It velocity, which is a vector and must be specified in magnitude and direction, is therefore $8$ m s$^{-1}$, south–west.

## SAQ 28

The vectors are as shown in Figure 50. One is at 20° to the direction of the roadway and has a length 6/5 times the length of the vector for the other car.

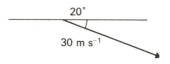

*Figure 50    The ratio of the lengths of the vectors is equal to the ratio of the speeds of the cars.*

## SAQ 29

Referring to Figure 41, the angle $\theta$ will be given by $\tan \theta = v_y/v_x$. In this case therefore $\tan \theta = -0.3/0.2 = -1.5$ and $\theta = -56°$ (to the nearest degree). The vector is at an angle of 56° below the $x$-axis (Figure 51). The magnitude of the vector is given by

$$v^2 = v_x^2 + v_y^2$$

Therefore

$v = \sqrt{[0.2^2 + (-0.3)^2]}$ mm s$^{-1}$

$= \sqrt{(0.04 + 0.09)}$ mm s$^{-1}$

$= \sqrt{(0.13)}$ mm s$^{-1}$

$= 0.361$ mm s$^{-1}$

The velocity is therefore compatible with the maximum cutting rate.

## SAQ 30

Since the car moves in a straight line, the velocity changes in magnitude, but not in direction. The initial velocity $v_i$ is smaller than the final velocity $v_f$ and the two are compared in Figure 52(a). The vector $v$ represents the velocity that must be added to $v_i$ to make it equal to $v_f$.

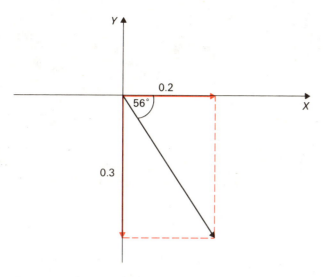

*Figure 51   The angle whose tangent is 1.5, is 56°; the angle whose tangent is −1.5, is −56° (tan (−θ) = −tan θ (Unit 4)). This is an angle in the fourth quadrant, that is below the positive X-axis.*

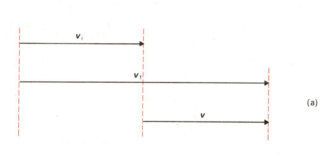

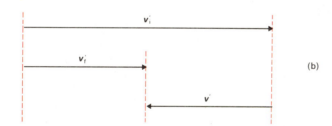

*Figure 52   (a) $v_i + v = v_f$. (b) $v_i' − v' = v_f'$.*

The acceleration is thus

$$a = (v_f − v_i)/t = v/t.$$

For this particular case

$$a = (24 − 12)/7 \, \text{m s}^{-2}$$
$$= 12/7 \, \text{m s}^{-2}$$
$$= 1.71 \, \text{m s}^{-2}.$$

The direction of $a$ is the direction of $v$, which is the direction of motion of the car.

When the car slows, the velocities can be represented by $v_i'$, $v_f'$ and $v'$ (Figure 52(b)).

The acceleration is

$$a = (12 − 24)/7 \, \text{m s}^{-2} = −1.71 \, \text{m s}^{-2}$$

in a direction opposite to the direction of motion of the car.

### SAQ 31

Figure 53 shows the vectors for the two forces drawn from the same point with an angle of 45° between them. Using the nose-to-tail additional rule the resultant force $r$ lies in a direction mid-way between the individual forces. The magnitude of $r$ is $r = 2f \cos 22\frac{1}{2}°$.

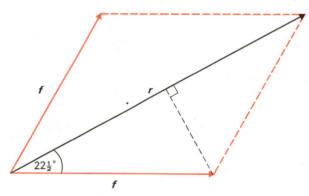

*Figure 53*

### SAQ 32

The problem can be solved by drawing the usual triangle of vector addition. An alternative is to add the forces together by components. I shall take horizontal and vertical components.

The one man exerts only a horizontal force on the trolley and its magnitude is 400 newtons. The other man exerts a force having both horizontal and vertical components. The magnitude of the vertical component is 500 sin 30° newtons and the magnitude of the horizontal component is 500 cos 30° newtons.

The horizontal component of the total force is thus 400 + 500 cos 30° = 833 newtons and the vertical component is 500 sin 30° = 250 newtons. This total force makes an angle with the horizontal whose tangent is 250/833 = 17° (to the nearest degree). Its magnitude is $\sqrt{(250^2 + 833^2)} = 870$ newtons.

59

# 6. Movement

# CONTENTS

## AIMS

The aims of this unit are:

1  To introduce Newton's Laws of Motion and the equations of motion in a straight line with constant acceleration.

2  To show how these can be used to model situations in which forces can be regarded as constant.

3  To teach how quadratic equations can be solved.

4  To introduce complex numbers.

## OBJECTIVES

When you have finished this unit you should be able to:

1  Explain mass, weight, momentum and gravitation and carry out simple calculations involving them (SAQs 3, 4, 5 and 6).

2  State and discuss Newton's Laws of Motion.

3  Identify the forces acting in given situations with a view to representing the significant ones in simple models (SAQs 1, 28 and 34).

4  Derive and apply the equations of motion in a straight line under constant acceleration (SAQs 7, 8, 9, 13, 14, 15, 16, 17, 32 and 33).

5  Use Newton's laws with the equation of motion to calculate distances, velocities and acceleration in particular situations involving one or two dimensions (SAQ 34).

6  Recognize a quadratic equation and be able to solve it, either by the formula method or by completing the square, and obtain either real or complex roots as appropriate (SAQs 18, 19, 20, 21 and 22).

7  Calculate the modulus of a complex number, give its complex conjugate and represent it on an Argand diagram (SAQs 23 and 24).

## STUDY GUIDE

Associated with Unit 6 *Movement*, are the television programme, *Crashing with Safety*, assignment material and optional self-test exercises on the computer. Full details of all these items can be found in the supplementary material mailed with Units 4–6.

Both the correspondence text and the television programme draw on models from the general theme of 'Road Safety'. If you have not already read the relevant section of *Modelling Themes* you should do so before continuing your work on Unit 6.

This text also draws on work covered in Block 1 and in Unit 5. From Block 1 you are expected to be familiar with the manipulation of algebraic equations and the use of symbols to represent either numbers or dimensioned quantities. From Unit 5 you will need to be familiar with the ideas of velocity, acceleration and force and with the concept of a vector. You will also need to be familiar with the shape of a parabola and the form of the equation which represents it.

After you have studied the unit you can use the objectives and summary as checklists of what you are expected to be able to accomplish.

# 1 INTRODUCTION

Two thousand five hundred years ago, or thereabouts, the Greeks were troubled with the idea of 'motion' or 'movement'. To them it was clear that two kinds of motion were understandable and natural, but a third required some special explanation. It was as obvious to them as it is to us that if you push something moveable, a cart, chariot, ship or plough, it will move, and that to keep it moving something has to keep pushing it—or pulling it—whether it be a man, a horse or the wind. It was also clear that it was natural for things to fall downwards towards the ground, for they regarded the ground as the natural place for all material things to rest, until someone picks them up. (They thought that fire, however, naturally rose upwards to the sky and found its way to the sun and the stars.) The kind of movement which they had difficulty in explaining was the flight of an arrow or a javelin. Nothing seemed to be pushing it along and yet for a great part of its flight it was not falling. They believed that such things must be driven on through the air by invisible 'movers' which had some strange power to move themselves and keep missiles in flight. The argument was extended to become a proof of the existence of God, because the power that kept the movers in motion must itself be unmoved and all powerful—and this must be God.

So there were three basic ideas about motion in ancient Greece. First that it was natural for things to move if a force was exerted upon them, secondly, that it was natural for things to fall, and, thirdly, that if something were moving other than downwards it must be moved by a 'mover'.

The Greeks also knew, of course, that the harder you push something the faster it moves, and they could feel, just as we can, that the downward force on heavy objects—due to their natural tendency to fall—feels greater for heavy objects than for light ones. Indeed, they knew that the weight or heaviness of an object is simply the downward force upon it. From this they concluded that heavy bodies fell faster than light ones. Very light ones like leaves and feathers could be seen to fall much more slowly than rocks or logs, in full accord with this point of view. That leaves are blown about in winter is a consequence of the well known invisible 'mover', the wind, acting upon them, as it does upon ships.

The inadequacy of this general concept of movement only emerged in the sixteenth century through the work of Galileo; its point of weakness lay in its main quantitative statement—that heavy bodies fall faster than light ones. (The idea that mathematics might be allied to physics was another enormous leap forward in human thought contributed by Galileo.) He demonstrated that two heavy objects, one twice the weight of the other, fell at the *same* speed. It is said that he dropped them both from the top of a tower—perhaps the tower of Pisa which was already conveniently leaning in Galileo's time. The two objects struck the ground together showing that they fell at the same speed. However, even such a demonstration was not wholly convincing for could not one of the objects have been aided by a 'mover'? So he also established his point by argument. He posed the question 'Will two equal weights, dropped separately, fall more rapidly than when they are tied together?' Two equal weights tied together form one object of double the weight of each, so according to Greek ideas this should fall twice as fast. But we know this cannot be: how could a small piece of cord make such a difference? The orthodox view was both wrong and unreasonable. Indeed, the idea of the interplay of experiment and reason, which characterizes science today, was greatly advanced by Galileo.

You can get some idea of the consternation caused by these discoveries of Galileo (which contradicted the views held by experts for over 1500 years) by noting the reaction of most of us to the achievements of Uri Geller in bending forks and other metal rods. We immediately think there must be some trick. Perhaps someday Geller's 'tricks' will be seen as a consequence of a better understanding of the natural world, but not yet. For a while, perhaps for ever, Uri Geller will be thought of simply as a clever conjurer. Galileo, however, won his case. His trick of dropping cannon balls from the top of a tower took us forward into the modern interpretation of motion which it is the first purpose of this unit to describe. They are enshrined in what we call Newton's Laws of Motion, although there is no doubt that the first one was due to Galileo.

# 2 NEWTON'S LAWS OF MOTION

## 2.1 Newton's First Law

The change that Galileo introduced is simply that he regarded it as natural, not only for a body to lie stationary on the ground, or to fall to the ground, but also that, once it has started to move, *it will continue to move in a straight line at a uniform speed without any help*. This idea, if you accept it, removes, at a stroke, all the 'movers' of the ancient Greeks. No longer was it *necessary* to conceive of a force acting on an object to keep it moving, it will move at uniform speed in a straight line without any help until a force comes to stop it or change its direction or slow it down or speed it up.

The flight of a javelin can now be seen as a combination of its tendency to fly in the direction in which it was thrown (influenced by the downward force which tends to make it fall) and the effect of the air tending to slow it down. Instead of looking for why it keeps going one looks for reasons why it does not!

Newton's First Law simply states this new view of the way things move.

*A body continues in a state of rest or uniform motion in a straight line unless acted upon by an external force.*

no horizontal force

*Figure 1    A situation in which Newton's First Law is exemplified.*

Newton called this a 'Law' because he believed that God made laws which everyone had to obey and that all he (Newton) had done was to discover some of them. Scientists now hold the view that Newton's First Law is only true for bodies which are neither very small nor travelling too fast. Quantum theory and relativity are more accurate models of reality than Newton's description and so it is best to think of Newton's First Law as a model of these more accurate theories. For all the problems which you will encounter in this course, however, Newton's model can be relied on, and it will be used in dealing with the relationships between forces and motion.

## 2.2 Newton's Second Law

Newton's First Law tells us that changes of motion are always due to forces. Newton's Second Law gives the relationship between the magnitude of a force and the acceleration that results from it.

In its simplest form Newton's Second Law states that

$$\text{acceleration of a body} = \frac{\text{force exerted on the body}}{\text{mass of the body}} \qquad (1)$$

or

$$\text{force} = \text{mass} \times \text{acceleration}.$$

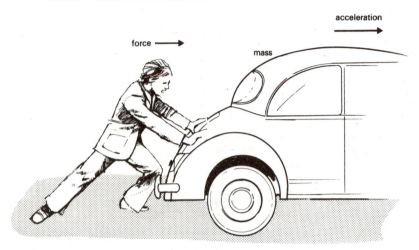

*Figure 2   An illustration of Newton's Second Law.*

This law applies to the common situation in which mass is constant, and, in particular, enables you to define units of force in terms of units of mass and acceleration. The SI unit of force is the *newton* (N). A newton is the force which will give a mass of one kilogram an acceleration of one metre per second in each second ($1 \, \text{m} \, \text{s}^{-2}$).    **newton**

A particular instance of the first form of this equation is therefore

$$1 \text{ metre per second squared} = \frac{1 \text{ newton}}{1 \text{ kilogram}}$$

Since you can measure acceleration by means of a tape measure and a watch, and you can measure the mass of an object by comparing it on a balance with a set of standard weights, you could use this equation to find how big a force one newton is. You have to find what force will give a mass of one kilogram an acceleration of $1 \, \text{m} \, \text{s}^{-2}$. A convenient way of imagining how big a force it is, is to think of the weight of an apple. The force you have to exert to hold up an apple is about 1 newton (1 N). If you were to arrange a small trolley, of mass 0.9 kg, so that it could run freely on a horizontal table and be pulled along by a suspended apple of mass 0.1 kg it will accelerate at a rate of about $1 \, \text{m} \, \text{s}^{-2}$.

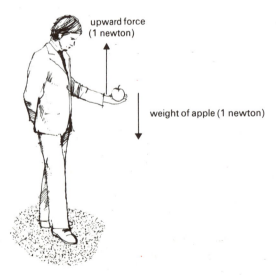

*Figure 3   Illustrating the force of 1 newton.*

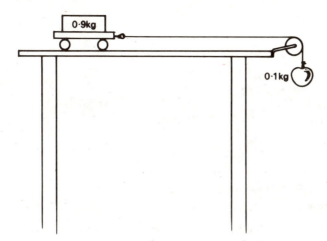

*Figure 4    The weight of a typical apple is about 1 newton. If this downward force causes a mass of 1 kg (the mass of the trolley plus the mass of the apple) to accelerate the acceleration will be about 1 m s$^{-2}$.*

It is said that Newton conceived the idea of gravitation by wondering why an apple fell to the ground when the moon, which he could see just behind the apple, did not fall. This legend, which I shall return.to in a moment, will perhaps help you to remember the approximate magnitude of the force of 1 newton.

*Example of the use of Newton's Second Law*

A Jumbo Jet aircraft has a mass of about 140 000 kg. Its four engines produce a thrust (i.e. a force) of about 800 000 N. If you ignore all other forces like friction with the runway, air resistance, etc., what acceleration will the aircraft experience at take-off?

Newton's Second Law tells us that

$$\text{acceleration} = \frac{\text{force}}{\text{mass}} \tag{1}$$

so in this case

$$\text{acceleration} = \frac{800\,000}{140\,000} \text{ m s}^{-2} = 5.7 \text{ m s}^{-2}$$

This calculation supposes that the force is constant, and that the mass is constant despite the burning of fuel.

**SAQ 1**

SAQ 1

Making similar suppositions calculate the forward thrust, via the wheels, produced by the Citroen and the Volvo.*

(a)   The mass of a Citroen Safari CX 2000 is 1320 kg. Its acceleration (presumed constant) from 30 to 50 mph (i.e. about 13 m s$^{-1}$ to 23 m s$^{-1}$) in top gear is 0.65 m s$^{-2}$.

(b)   The mass of a Volvo 245DL is 1280 kg and it accelerates in top gear from a velocity of 13 m s$^{-1}$ to 23 m s$^{-1}$ in 9.7 s.
List the suppositions you make in this calculation.

(c)   The mass of a Ford Granada 3000 GL is 1400 kg and it accelerates from 13 m s$^{-1}$ to 23 m s$^{-1}$ using automatic transmission in 4.4 s, (i.e. it changes gear automatically). Which of your suppositions may no longer be appropriate?

* Motor, *July 1976*

9

## 2.3 Mass, weight and gravitation

Newton's Second Law of Motion has introduced the concept of *mass*, a term you may not be familiar with for it is not easy to define. It is usually defined as 'the quantity of matter a body possesses', but this is not very informative. It is customary to speak about the weight of an object even when we mean mass, so let me try to explain the difference between mass and weight.

mass

That there is a difference of some kind must be clear to you in this space age when astronauts weigh less on the Moon than they do on Earth and when in an orbiting satellite they are even weightless. Although their weight changes, their mass (i.e. the quantity of matter in them) does not change significantly—only as a result of eating, drinking, perspiring, slimming, etc.

Let me go back to the discovery of the force of *gravitation*. Galileo, who died in the year that Newton was born, had no conception of gravitation: indeed, he believed that planetary orbits must be circular in order to preserve the perfection of the heavens. However, as well as discovering that all heavy bodies fall at the same rate—and perhaps lighter ones too if there were no air resistance—Galileo discovered that they all fell with *constant* acceleration. He expressed this in a rather neat way. He postulated that the distances travelled during equal intervals of time by a body falling from rest (i.e. having zero initial velocity) are in proportion to the odd numbers. Thus, if a cannon ball falls 1 m in the first interval of time it will fall 3 m in the next interval, 5 m in the third interval, and so on.

gravitation

### SAQ 2

SAQ 2

Complete the table below of distances covered in equal intervals of time for a cannon ball rolling downhill and so infer the total distance covered after $T$ intervals of time (see Figure 5.)

| Time/ number of intervals | Distance covered in corresponding interval/m | Total distance covered/m |
|---|---|---|
| 1 | 1 | 1 |
| 2 | 3 | $1 + 3 = 4$ |
| 3 | 5 | $1 + 3 + 5 = 9$ |
| 4 | 7 | $=$ |
| 5 | 9 | $=$ |
| 6 | 11 | $=$ |
| $T$ | | |

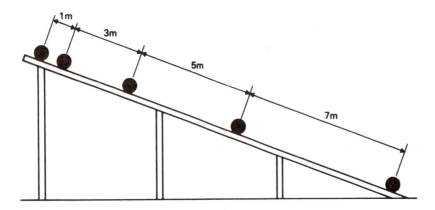

*Figure 5   A heavy ball rolling down an inclined plane. The distances shown are covered in equal intervals of time.*

Actually, Galileo had no accurate clocks; he used his own pulse to measure intervals of time and so had to reduce the acceleration he was measuring by rolling balls down inclined planes rather than by just dropping them.

However, he inferred that falling bodies also fall with constant acceleration. This discovery by Galileo of constant acceleration was simply a 'rule of thumb'. It offered no explanation of why bodies all fell with equal, constant acceleration; it simply recorded the fact that they do. Newton's contribution was to add the idea that not only is each body near the surface of the Earth attracted to it by a force proportional to its mass, but that *all* bodies attract each other by a 'gravitational' force. His moment of inspiration is said to have been when he realized that this force which causes an apple to fall could be the same one which causes the Moon not to fall but to orbit the Earth. In the case of the Moon, he argued, the force changes the Moon's velocity by changing *its direction* without changing its speed, because the acceleration is at right-angles to the Moon's velocity. For the falling apple acceleration and velocity are in the same direction.

So according to Newton's theory of gravitation the weight of an object, which you feel when you lift it, is the force of attraction between the Earth and that object.

How is it then that this force of gravity leads to the phenomenon of constant acceleration that Galileo discovered? Newton's gravitational theory states that for two bodies with masses $m_1$ and $m_2$, the gravitational force $f$ between the bodies is

$$f = \frac{\text{constant} \times m_1 \times m_2}{(\text{distance between them})^2}$$

or

$$f = \frac{Gm_1m_2}{r^2} \tag{2}$$

$G$ is called the *gravitational constant* and can be measured. (The capital letter $G$ is universally used to denote the gravitational constant and so I have used it here even though it contravenes the course convention where lower case letters should be used to stand for dimensioned quantities.)

gravitational constant

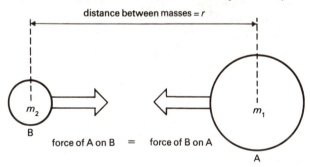

*Figure 6  Gravitational forces. Each mass attracts the other with a force given by:* $\text{constant} \times m_1m_2/r^2$

Now if we consider the forces on two objects, say a brick and a stone, of masses $p$ and $q$ near the surface of the Earth then, in the gravitation equation, the gravitational constant $G$, the mass of the Earth $m_1$ and the distance $r$ to the centre of the Earth will be the same or nearly so for each mass. So the equations can be written as

$$\text{force on the brick} = \frac{Gm_1p}{r^2} = \left(\frac{Gm_1}{r^2}\right)p \tag{3}$$

$$\text{force on the stone} = \left(\frac{Gm_1}{r^2}\right)q \tag{4}$$

The force in each case is a constant term $(Gm_1/r^2)$ multiplied by the mass. By Newton's Second Law, acceleration is the term which relates force to mass; so if we call this acceleration $g$ 'the acceleration due to gravity', we can write equations 3 and 4 as

$$\text{force on the brick} = gp$$

$$\text{force on the stone} = gq$$

where the acceleration of each mass is the same and is given by

$$g = Gm_1/r^2 \tag{5}$$

11

Thus, Newton's theory explains why all bodies fall under gravity (if air resistance is negligible) with the same acceleration. It also explains why you can use a balance to compare masses of bodies even though a balance compares weights, which are downwards forces due to gravitation. Provided the distance $r$ for each body is the same (i.e. the arms of the balance are not so long that the distance of one body to the centre of the Earth differs significantly from the distance of the other from the centre of the Earth) the term $Gm_1/r^2$ will be the same for each mass so the ratio of masses is the same as the ratio of the forces.

### SAQ 3

SAQ 3

Explain why the weight of an object on the Moon is less than the weight of the same object on Earth.

Nowadays, with electronic timers it is not difficult to measure very accurately the rate at which objects fall to the ground. It turns out that $g$, the acceleration downwards of all objects in Britain near the surface of the Earth, if air resistance is negligible, is 9.81 metres per second squared to three significant figures (i.e. $9.81 \, \text{m s}^{-2}$). But because the Earth is not quite spherical the value of $g$ is not the same at all parts of the Earth. Because the distance $r$ to the centre of the Earth is shorter at the poles than at the equator the value of $g$ is greater there; about $9.83 \, \text{m s}^{-2}$ as compared with $9.78 \, \text{m s}^{-2}$ at the equator.

Would you expect $g$ to be (a) the same (b) greater (c) less at the top of Mount Everest as compared with its value at the base?

The answer is (c). $g = \dfrac{Gm}{r^2}$, and at the top of Mount Everest $r$ is *larger* (i.e. the distance to the centre of the Earth is larger) and therefore $g$ is *smaller*.

### SAQ 4

Use $g = 9.81 \, \text{m s}^{-2}$ to calculate the mass $m_1$ of the Earth given that the gravitational constant $G$ has the value of $6.67 \times 10^{-11} \, \text{m}^3 \, \text{kg}^{-1} \, \text{s}^{-2}$ and that the radius of the Earth is about 6370 km.

*Method*

You know that

(a)   the force on an object with mass $m_2 = \dfrac{Gm_1 m_2}{r^2}$ and

(b)   for this mass $m_2$ near the surface of the Earth

$$\text{the acceleration } (g) = \frac{\text{force on } m_2}{m_2}$$

(a) and (b) are simultaneous equations giving $m_1$.

What then is the difference between the mass of an object and its weight? *The mass is the quantity of matter it contains and the weight is the force of gravity acting upon it.*

But the downward force due to gravity on an object is $g$ times its mass, so *the weight of an object of mass m is mg, that is (9.81 × m) newtons in SI units.*

The weight of one apple is a force of about 1 N, because there are about 10 typical apples to the kilogram.

Weight is a force you can feel, but mass is *not* something you feel or perceive directly. You can determine the mass of an object from its weight in circumstances where gravitation acts, or from the impact of it hitting you, as I will discuss in a moment.

Given that the mass of the Moon is $7.34 \times 10^{22}$ kg and its radius is 1740 km, calculate the weight on the Moon of a body with a mass $m$.

It often happens that forces, like the thrust of a jet engine or the drive of a car, are given in kilograms—when they should really be given in newtons. When this is done it means that the force is equivalent to the weight of so many kilograms. Thus, if the thrust of Jumbo Jet aircraft engines is given as 80 000 kg it means that they exert a force of nearly 800 000 N (i.e. 80 000 × 9.81 N).

## 2.4   Momentum

You may recall that I introduced Newton's Second Law of Motion in its simplest form as

force = mass × acceleration

The full expression of Newton's Second Law is that *the rate of change of momentum of an object is equal to the external force acting upon it.*

*Momentum* is the name given to the product of the mass $m$ and the velocity $v$ of an object. So

momentum = mass  × velocity

$$= mv \tag{6}$$

If the mass remains constant during the time the force is acting then the *rate of change of momentum* becomes *the mass times the rate of change of velocity.* But rate of change of velocity is acceleration. For this case Newton's Second Law reduces to

force = mass × acceleration

This is the simplified form we have been discussing until now.

The full expression of Newton's Second Law is

force = rate of change of momentum       (7)

Evidently the full expression of Newton's Second Law becomes important when you are dealing with objects whose masses change with time—like rockets or aircraft which burn fuel as they fly, so that their masses decrease significantly in flight. For a constant force the acceleration gets bigger as the mass decreases. This is why big space rockets seem to begin to accelerate rather slowly. As they get lighter they accelerate more quickly. (Also, as they rise higher the force of gravity gets smaller.)

Momentum is an important quantity for calculating the velocities of colliding bodies, because the total momentum of the bodies is the same before and after the collision if there are no external forces acting.

Momentum is a measure of *impact*. A small bullet travelling very fast or a bulldozer travelling slowly could both hit you very hard if you got in their way; the former because of its large velocity, the latter because of its large mass.

When you hit a golf ball the head of the golf club slows down at impact as the golf ball takes off. The loss of momentum of the golf club is equal to the gain of momentum of the golf ball. This is because the total momentum before collision equals the total momentum after collision.

The momentum before impact is

mass of golf club head times its velocity

momentum

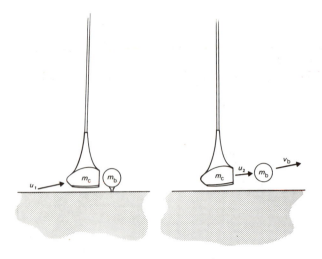

*Figure 7  Impact and the conservation of momentum.*

$$m_c u_1 = m_c u_2 + m_b v_b$$

*where $m_c$ is the mass of the golf-club head, $m_b$ is the mass of the ball, $u_1$ and $u_2$ are the velocities of the golf-club head before and after impact, respectively, and $v_b$ is the velocity of the ball after impact.*

The momentum after impact is

> mass of golf club head times its final (slower) velocity

> +

> mass of golf ball times the ball's initial velocity

The momentum before impact must equal the momentum after impact.

Colliding objects do not have to fly apart again for this relationship to be true. If a bullet hits a stationary man and lodges in him the momentum of the bullet before impact will be the same as the momentum of the two together after impact.

*Example*

A car of mass 700 kg travelling at 30 m s$^{-1}$ collides with a stationary van of mass 1300 kg, and after the collision they remain locked together. At what speed, immediately after impact, do the two together skid forward? Imagine they are on a slippery road if you like.

Let the speed at which they move off be $V$ m s$^{-1}$

Total momentum of vehicles before impact $= 700 \times 30$ kg m s$^{-1}$

Total momentum of vehicles after impact $= (700 + 1300)V$ kg m s$^{-1}$

But these two quantities are equal,

so        $700 \times 30 = 2000V$

$$V = \frac{21\,000}{2000} = 10.5$$

and the new speed is 10.5 m s$^{-1}$.

**SAQ 6**

Some Western films show the cowboys being knocked backwards when they are supposed to have been shot. Others show them falling forwards or just collapsing. If a bullet of mass 20 g is fired at a speed of 0.5 km s$^{-1}$ and is supposed to hit and lodge in a man whose mass is 50 kg, which film director is right? Would you expect to see the man knocked violently backwards or only slightly?

## 2.5  Newton's Third Law

You can understand why momentum is neither lost nor gained at impact by considering Newton's Third Law. This law states that:

*action and reaction are equal and opposite.*

Action means the exertion of a force in one direction; reaction means a counter exertion in the opposite direction. So, if you push against a wall you will experience a force equal to the force of your push. If the wall is the side of a swimming bath your pushing it will drive you away from it. The weight of a body *at rest* (which is a downwards force) is equal and opposite to the upwards force (the reaction) of the table or chair supporting it. When a bullet hits a man he exerts a force on it slowing it down or stopping it. When two cars collide they exert equal and opposite forces on each other.

The implications of this law are somewhat more significant or surprising in connection with rocket flight. A propeller-driven aircraft moves forward by using its propeller to push backwards on the air. Similarly, the screw of a ship pushes backwards on the water and the reaction of the water pushes the ship forward. It is tempting then to suppose that a rocket pushes itself forward in a similar manner by pushing on the air with its jet of hot gas and thereby being pushed forward by the 'reaction' of the air. But rockets work satisfactorily in the nearly total vacuum of outer space—indeed they work better beyond the constraining resistance of the Earth's atmosphere. How do they work?

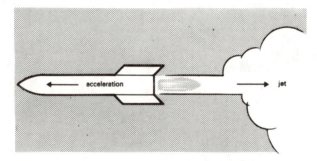

*Figure 8  Rocket propulsion. Change of jet momentum equals change of momentum of rocket.*

Simply by ejecting gas backwards (action) the rocket is propelled forwards (the reaction). A small amount of gas ejected at high speed produces a movement of the rocket in the opposite direction—regardless of whether or not there is any air surrounding the rocket. Indeed, the momentum imparted to the jet of hot gases is equal and opposite to the momentum given to the rocket.

You could have answered SAQ 6 (about whether a cowboy is knocked backwards by a bullet) by noting that the kick he receives from being shot cannot be bigger than the kick experienced by the man who shot him. The gunman with a rifle at his shoulder experiences a reaction equal to the action of accelerating the bullet down the barrel of the rifle. The bullet loses a little speed in flight so it cannot do more than impart a somewhat smaller momentum to the man the bullet hits.

Why is momentum neither lost nor gained at impact?

Because action and reaction are equal and opposite two objects in collision exert equal and opposite forces on each other. By Newton's Second Law they

15

inflict equal and opposite rates of change of momentum upon each other. Because they must experience these equal and opposite forces for equal times, it follows that the *total* change of momentum of each must be the same. Hence there is no gain or loss of momentum at impact. Equally, there is no net gain or loss of momentum when things fly apart, like a gun and its bullet or a rocket and its jet.

Returning for a moment to Newton's First Law, perhaps you can now see a problem. It seems that a body *can* change its 'state of rest or uniform motion in a straight line' without an *external* force acting by simply ejecting some of itself. It will then recoil with a new velocity in some direction or other. The answer is that 'a body' is not, in Newton's First Law, divisible in this way. You must continue to think only of the entire body, see Figure 9. If it flies apart,

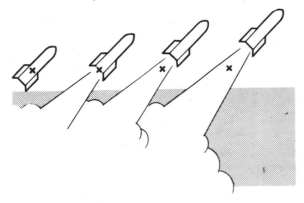

*Figure 9   The crosses show the centre of gravity of a rocket moving in space at equal intervals of time. Even though the rocket does not keep to a straight path the centre of gravity of the rocket and its exhaust does.*

some going one way and some the other, its centre of mass carries on unaffected. The centre of mass, or centre of gravity as it is sometimes called, is the point about which an object can, if rigid, be supported by a single force without any rotation. It's at the centre of a wheel or the middle of a uniform beam. You can calculate where it is even if the object is not rigid, so a rocket and its jet of hot gases have a centre of mass too, and this centre of mass of a rocket *and* its jet (in a vacuum) does *not* change direction. The rocket and jet fly in opposite directions and if you take both into account Newton's First Law is not violated—the centre of mass continues to move at uniform velocity unless acted upon by an external force.

### 2.6   Summary

1   Newton's Laws of Motion:

*Newton's First Law*
A body continues in a state of rest or uniform motion in a straight line unless acted upon by an external force.

*Newton's Second Law*
The force exerted upon a body is equal to the rate of change of its momentum, provided $m$ remains constant.

      force = mass × acceleration

      $f = ma$

*Newton's Third Law*
Action and reaction are equal and opposite.

2   Momentum is the product of mass and velocity.

      momentum $= mv$

Momentum (of movable objects) is conserved at impact. That is, the total momentum before impact is the same as the total momentum after impact providing no external forces act on the objects.

3   Newton's Law of Gravitation states that two bodies with masses $m_1, m_2$ a distance $r$ apart attract one another with a force given by

$$\text{force} = \frac{Gm_1m_2}{r^2}$$

where $G$ is a dimensioned constant known as the gravitational constant.

4   For a body of mass $m_2$ near the surface of the Earth the force of gravity is given by

$$\text{force on mass } m_2 = \frac{G \times \text{mass of the Earth}}{(\text{radius of the Earth})^2} m_2$$

Hence, by Newton's Second Law, the acceleration due to gravity $g$ is given by

$$g = \frac{G \times \text{mass of the Earth}}{(\text{radius of the Earth})^2}$$

In SI units $g = 9.81 \, \text{m s}^{-2}$, $G = 6.67 \times 10^{-11} \, \text{m}^3 \, \text{kg}^{-1} \, \text{s}^{-2}$

# 3  MOTION IN A STRAIGHT LINE WITH CONSTANT ACCELERATION

The motion of a good many bodies (for example, cars, falling objects, trains, missiles) can be modelled on the supposition that, for part of their journey at least, they have constant acceleration. With this simplification it is possible to obtain a number of general equations which can be used for calculations of distance covered, final velocity, etc.: for example, during an aircraft take-off, or when a car accelerates.

Before going on to consider these equations I want to remind you of a course convention you met in Unit 3. In this convention, lower-case letters are used to represent dimensioned quantities and capital letters are used to represent numbers. An equation is written either wholly in dimensioned quantities or wholly in numbers. However, sometimes we may wish to substitute numerical values for some, but not all, of the variables or parameters in a dimensioned equation. How can this be done?

For instance, suppose an object of mass $m_1$ moving at velocity $u$ collides with a stationary object of mass $m_2$ and the two objects stick together and move off at a velocity $v$. Then from Section 2.4 you will see that

$$\text{initial momentum} = m_1 u$$

$$\text{final momentum}\ \ = m_1 v + m_2 v$$

so that

$$m_1 u = m_1 v + m_2 v. \tag{10}$$

Now if it is known that

$$m_1 = 20\,\text{kg}$$

$$m_2 = 5\,\text{kg}$$

$$u = 10\,\text{m s}^{-1}$$

and it is required to find $v$, then we want to put numerical values for $m_1$, $m_2$ and $u$ into equation 10. There is no numerical value for $v$, but in order to write down an equation in terms of $v$ we can write

$$v = V\text{m s}^{-1}$$

$V$ represents a number—the number of metres per second at which the two objects move away.

Equation 10 can now be written as

$$20 \times 10 = 20V + 5V$$

which means, solving the equation, that $V = 8$. Since $v = V\text{m s}^{-1}$, this means that $v$, the velocity at which the two objects move, is $8\,\text{m s}^{-1}$.

It is important for you to realize that this convention is *only* adopted by this course, and that it has been done in the hope that it will help you to be clear about whether an equation is in terms of numbers or of dimensioned quantities. You have already met a case where the course team have had to deviate from their own rules in that they have agreed to use $G$, the universally accepted symbol for the dimensioned quantity called the gravitational constant! After this course, you will be expected to adopt the normal practice of determining for yourself whether a symbol represents a number or a dimensioned quantity from the context in which it is used.

## 3.1  General equations of motion

Here we are concerned with motion in which the acceleration $a$ is constant. The acceleration due to gravity $g$—which is almost constant—is a particular value of $a$ applicable to bodies falling in a vacuum or when air resistance is negligible.

If the acceleration $a$ lasts for a time $t$ the increase in velocity is $at$. But this increase is the difference between the final velocity $v$ and the initial velocity $u$. So if we call the initial velocity $u$ and the final velocity $v$

$$v - u = at \tag{11}$$

The *average* velocity over this period is $\frac{1}{2}(v + u)$ because acceleration is constant, so the distance travelled $s$ is the average velocity $\frac{1}{2}(v + u)$ multiplied by the time $t$. So

$$s = \frac{1}{2}(u + v)t \tag{12}$$

Using this equation we can calculate how far a car goes when it accelerates uniformly from $14\,\mathrm{m\,s^{-1}}$ to $24\,\mathrm{m\,s^{-1}}$ in $4.4\,\mathrm{s}$. In equation 12, $u = 14\,\mathrm{m\,s^{-1}}$, $v = 24\,\mathrm{m\,s^{-1}}$, $t = 4.4\,\mathrm{s}$ and $s$ (the distance covered) $= S\,\mathrm{m}$.

$$S = \frac{(14 + 24)}{2} \times 4.4 = 83.6$$

The distance covered is therefore $83.6\,\mathrm{m}$.

Sometimes it is more convenient to have an equation which does not involve $v$. From equations 11 and 12 we can eliminate $v$ by substituting for $v$ from equation 11 into equation 12. It we add $u$ to both sides of equation 11 we get

$$v = u + at$$

Substituting for $v$ in equation 12 we get

$$s = \frac{1}{2}(u + u + at)t$$

$$s = \frac{t}{2}(2u + at)$$

or $\qquad s = ut + \frac{1}{2}at^2 \tag{13}$

*Example*

If a Jumbo Jet takes $30\,\mathrm{s}$ to take off and it accelerates at $5.7\,\mathrm{m\,s^{-2}}$, as calculated earlier, then the distance it travels, starting from rest ($u = 0$) is $S$ metres, where

$$S = 0 + \frac{1}{2} \times 5.7 \times 900 = 2565$$

On other occasions it is convenient to use an equation involving $v$, but not $t$.

### SAQ 7
SAQ 7

Show that by eliminating $t$ between equations 11 and 12 you obtain the equation

$$v^2 = u^2 + 2as \tag{14}$$

We now have four equations each relating four of the five variables $u, v, a, s$ and $t$. They are useful in carrying out calculations about the motion of bodies moving along straight lines with an acceleration which is constant at least for the duration of the period $t$. These equations are

$$v = u + at \tag{11}$$
$$s = \frac{1}{2}(u + v)t \tag{12}$$
$$s = ut + \frac{1}{2}at^2 \tag{13}$$
$$v^2 = u^2 + 2as \tag{14}$$

We will consider later in the course how to calculate distances, etc., when acceleration varies with time.

### SAQ 8
SAQ 8

Each of the above four equations omits one of the five variables. There must therefore be five such equations. Derive the equation which completes the set.

19

Most people who carry out the sort of calculations which use these equations usually only try to remember the above four equations. They are in your *Handbook* for easy reference.

## 3.2 Using Newton's Laws

*Example*

A ball is rolling downhill from rest at a constant acceleration of $5\,\mathrm{m\,s^{-2}}$.

How fast is it travelling after 3 s, and how far has it travelled?

To find out how fast it is travelling, we know the initial velocity $u$ is zero and the acceleration $a$ is $5\,\mathrm{m\,s^{-2}}$. We also know that $t = 3\,\mathrm{s}$ and we are asked to find $v$. The equation we need to relate $u$, $a$, $t$ and $v$ is equation 11,

$$v = u + at$$

and if we let $v = V\,\mathrm{m\,s^{-1}}$ then

$$V = 0 + 3 \times 5$$
$$= 15$$

and after 3 s the ball is travelling at $15\,\mathrm{m\,s^{-1}}$.

To find how far it has travelled, we again know $u = 0$, $a = 5\,\mathrm{m\,s^{-2}}$ and $t = 3\,\mathrm{s}$ and $s$ is the unknown so we need equation 13:

$$s = ut + \tfrac{1}{2}at^2.$$

Putting $s = S\,\mathrm{m}$,

$$S = 0 + \tfrac{1}{2} \times 5 \times 9$$
$$= 22.5$$

so the ball has rolled 22.5 m.

### SAQ 9

Look back at SAQ 2 on p. 10. If the chosen interval of time in SAQ 2 were 12 s what would be

(a)  the speed of the cannon ball after each interval of time?

(b)  the constant acceleration throughout the period of observation?

## Stopping distances of a car

In this section I am going to look first at two situations involving motion and show how to calculate quantities which may be of importance.

Have you read *The Highway Code*? If so you will be familiar with diagrams like that of Figure 10 and you may have wondered how these stopping distances were obtained. You will notice that there are two parts to each of the distances, the thinking distance and the braking distance. You can either try to measure them or try to calculate them. Here, I am going to show you how to model the situation, and use the equations we have obtained to carry out the calculations. The quantities shown in Figure 10 were calculated in the following way:

(a) *Thinking distance*

This is the distance the car travels after the driver has decided to stop and before his or her foot operates the brake pedal. How long would you say it takes—a second—or less? 0.6 s is about average. So let us suppose first that it always takes 0.6 s.

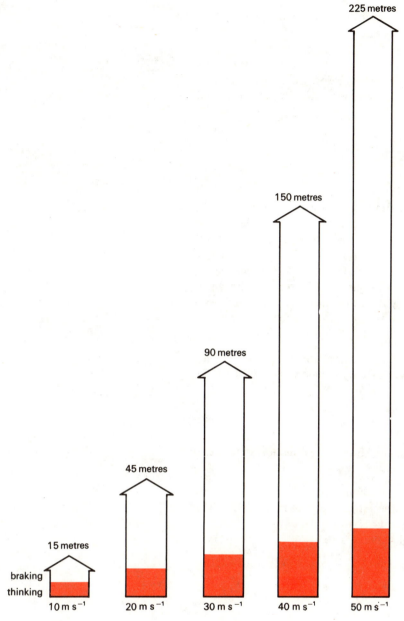

225 metres

150 metres

90 metres

45 metres

15 metres

braking
thinking

10 m s$^{-1}$    20 m s$^{-1}$    30 m s$^{-1}$    40 m s$^{-1}$    50 m s$^{-1}$

*Figure 10    Stopping distances (as given in* The Highway Code).

If the car's velocity is $u$ and the thinking time is $t_t$ then the car will have gone a distance $d_t = ut_t$ before the driver's foot hits the brake.

Using the SI system of units, I shall put $d_t = D_t$ m, $u = U$ m s$^{-1}$ and $t_t = T_t$ s, where $D_t$, $U$ and $T_t$ are numbers. $T_t = 0.6$ and so if $U = 1$, $D_t = 0.6$, if $U = 6$, $D_t = 3.6$, and so on. We can draw a graph and then read off the thinking

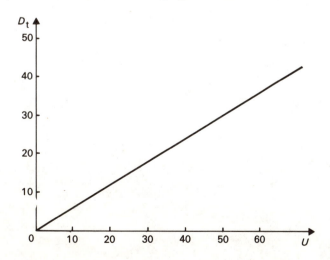

*Figure 11    Graph of the number of metres travelled D$_t$ while thinking against the number of metres per second U.*

distances for any given speed, as shown in Figure 11. You will notice that it is a linear (straight line) graph whose equation is

$$D_t = 0.6U \qquad (U \geqslant 0) \tag{15}$$

Remember $D_t$ is the number of metres travelled before the driver reacts and $U$ is the number of metres per second of speed. From the graph we can read off the thinking distances for any constant speed we like. At $44\,\mathrm{m\,s^{-1}}$ (100 mph) the thinking distance is nearly 30 metres (100 ft). Quite a long way to travel before your foot even gets to the brake!

### (b) Braking distance

Let us suppose that the total braking force $f$ for a given vehicle is the same at all speeds. (Actually of course the maximum braking force depends on the road surface, the state of the weather, the state of the tyres and whether the road is flat, but for the rough general conclusions we want here it is a good enough supposition.) If the car is travelling forwards with a velocity $u$ before it brakes, then on braking the force exerted will be in the backwards direction. To indicate that the force is in the opposite direction to $u$ it is convenient to say that the force is negative, thus distinguishing between a forward force (obtained when the driver's foot is on the accelerator rather than the brake) and a backward or retarding force.

If the mass of the car plus its contents and occupants is $m$ (and we assume they are reasonably firmly fixed in the car), then by Newton's Second Law the acceleration $a$ is given by

$$a = \frac{f}{m} \tag{16}$$

Because $f$ is negative the acceleration is in the opposite direction to the initial velocity and is therefore a retardation.

Obviously, different cars have different braking forces and different masses. But in the spirit of our earlier supposition (that the force is constant) let us suppose that the maximum braking force is two thirds of the weight of the vehicle. This is a reasonably typical value.

Thus, since $mg$ is the weight of the vehicle, $f = -2mg/3$. So, using equation 16

$$a = \frac{-2mg}{3m} = -\frac{2}{3}g$$

But $\qquad g = 9.81\,\mathrm{m\,s^{-2}}$

So $\qquad a = -6.5\,\mathrm{m\,s^{-2}}$.

The retardation is $6.5\,\mathrm{m\,s^{-2}}$

Now we want to know the braking distance $d_b$ for various values of initial velocity $u$. We know the final velocity $v$ is zero and we know the acceleration, so we want an equation relating $v$, $u$, $a$ and $s$; namely

$$v^2 = u^2 + 2as$$

On substituting $d_b$ for $s$ and putting $v = 0$, the equation reduces to

$$d_b = \frac{-u^2}{2a}$$

Putting in a numerical value for $a$, and remembering that this means that all the variables should be expressed as numbers, this gives

$$D_b = U^2/13, \text{ since } a = -6.5\,\mathrm{m\,s^{-2}}. \tag{17}$$

When $U$ is 10, $D_b$ is $\dfrac{10^2}{13} = 7.7$ (to two significant figures—use your slide rule to check this).

When $U = 20$, $D_b$ is $\dfrac{20^2}{13} = 31$, and so on.

If you draw a graph of $D_b$ against $U$ will you get a straight line?

No, you get a curve which is part of a parabola (see Unit 5).

**SAQ 10**

**SAQ 10**

Use your slide rule to complete the missing entries in this table.

| Initial velocity/m s$^{-1}$ $U$ | 0 | 5 | 10 | 20 | 30 | 40 | 50 |
|---|---|---|---|---|---|---|---|
| Braking distance/m $D_b$ | 0 | | 7.7 | 31 | | | |

Now plot a graph of $D_b$ against $U$ on the graph paper of Figure 12.

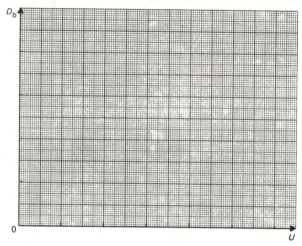

*Figure 12   Graph paper for SAQ 10.*

(c) *Stopping distances*

The total stopping distance is the thinking distance plus the braking distance. Here again it is convenient to adopt the capital letter convention and to adhere firmly to the SI system of units. Each capital letter is now a number standing for the number of each appropriate unit. At an initial velocity $U$ m s$^{-1}$ the thinking distance is $0.6U$ m, the braking distance $D_b = U^2/13$ m, so the total stopping distance $D_s$ m is given by

$$D_s = 0.6U + \frac{U^2}{13} \qquad (U \geqslant 0) \qquad (18)$$

This then is the equation relating the number of metres needed to stop a car from an initial velocity $U$ m s$^{-1}$. 0.6 s is the typical thinking time of a driver expressed in seconds.

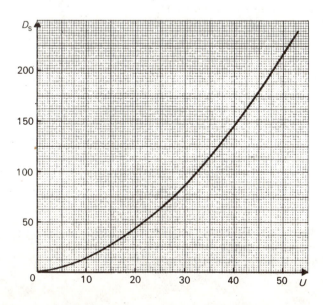

*Figure 13   Stopping distances against speed.*

23

The graph of $D_s$ against $U$ is given in Figure 13. It too is part of a parabola. From this graph you can read off the total stopping distances for $10\,\mathrm{m\,s^{-1}}$, $20\,\mathrm{m\,s^{-1}}$, $30\,\mathrm{m\,s^{-1}}$, ..., etc. and check them against those given in Figure 10.

This graph and Figure 10 give the minimum distances which it takes a vehicle to stop when travelling at a given velocity. Not all vehicles will take the same distance, on wet roads the braking distances may double, on icy roads they may be more than double. Some cars may not have as good brakes as they should. With a tired or drunken driver the thinking distances may increase substantially.

In using this model the Ministry of Transport has made many suppositions which they know do not apply to every vehicle, but they hope it will serve the purpose of giving motorists some idea of the magnitude of the minimum distances in which their vehicles can stop in good weather conditions.

These stopping distances can alternatively be found algebraically from equation 18

$$D_s = 0.6U + U^2/13 \qquad (U \geqslant 0)$$

by substituting in the different values of $U$. For instance, when $U = 10$

$$D_s = 0.6 \times 10 + 10^2/13$$
$$= 6 + 100/13$$
$$= 13.69$$
$$= 13.7 \text{ (to three significant figures)}$$

So the stopping distance from an initial velocity of $10\,\mathrm{m\,s^{-1}}$ is $13.7\,\mathrm{m}$.

### SAQ 11

SAQ 11

Substitute into equation 18 to find the stopping distances for velocities of $20\,\mathrm{m\,s^{-1}}$, $30\,\mathrm{m\,s^{-1}}$, $40\,\mathrm{m\,s^{-1}}$. Give your answer to three significant figures.

Why do you think that these do not agree exactly with those given in Figure 10?

### (d) Speed limits

There is another purpose which this model can serve, namely that of deciding what speed limits might be set before major road hazards like a roundabout or road junction or bend. There is a danger that drivers of cars travelling along a motorway at 70 mph may not realize how fast it is safe to go near a termination. Unless advised by speed limit signs, they may not slow down sufficiently to be able to stop at the terminating roundabout. The speed indicated should be such as to be safe for tired drivers on wet roads. In this case, the thinking distance is not significant because road signs give ample warning, but the braking distance may be doubled to allow a well controlled deceleration. Hence, the calculation should perhaps be done using double the braking distance. Figure 14 shows the graph of this distance $D$ metres against $U\,\mathrm{m\,s^{-1}}$, the initial velocity, where

$$D = \frac{2U^2}{13} \qquad (19)$$

At what distance from the roundabout should the first speed limit sign (60 mph) be placed?

The conversion factor from miles per hour to metres per second is $\dfrac{1}{2.25}$, so 60 mph is $\dfrac{60}{2.25}\,\mathrm{m\,s^{-1}}$ or $26.7\,\mathrm{m\,s^{-1}}$. The line $U = 26.7$ (dotted line on Figure 14) intersects the curve at about $110\,\mathrm{m}$. Because cars do not usually slow down to the speed limit until just after the sign it would probably be advisable to put the sign about 120 to $130\,\mathrm{m}$ before the roundabout. (*Note:* instead of reading $110\,\mathrm{m}$ off the graph I could have substituted $U = 26.7$ into equation 19.)

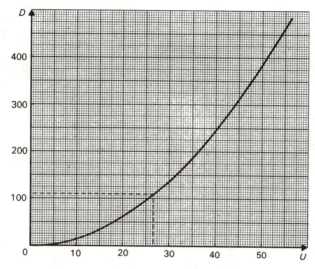

*Figure 14 Double the calculated braking distance as a function of speed. This allows for inaccuracy in supposing that all brakes are equally effective.*

### SAQ 12

SAQ 12

Assuming there is to be a speed limit sign of 40 mph before the roundabout, where should it be placed?

## Motion under the influence of gravity

Perhaps it is to be regretted that so many advances in science, mathematics and technology have been achieved as a result of attempts to improve military effectiveness, but it is certainly true that Galileo undertook his studies of the motion of projectiles and other falling bodies in order to improve the accuracy of guns. Presumably the Greeks 1800 years earlier had similar aims when they were developing their own theories to explain the flight of projectiles.

Early descriptions of longish range projectiles, as viewed from the starting point, were of missiles rising along an inclined path away from the viewer, and then just dropping vertically. The flight of a well hit golf ball looks a bit like this from the tee. This we now know is too simple a model to be very helpful for any practical purposes, but it does, in a simple way, separate vertical motion from the rest.

A crucial step in the achievement of our present understanding of the flight of projectiles was the resolution of their fairly complex flight into two directions, both upwards and forwards. The simplest model is one in which the horizontal component of a projectile's velocity is constant, while the vertical component is like a projectile moving vertically influenced by the constant force of gravity. Both are normally affected by air resistance, winds and even the spin of the missile, but a simple model which ignores these factors gives quite good results for low speeds and dense bodies such as Galileo's cannon balls.

I shall consider these two dimensions of motion separately, taking first vertical motion in what I shall call 'the *y*-direction' and then horizontal motion in 'the *x*-direction' and then combining them together.

### Vertical motion: falling

Air resistance has negligible effect upon most heavy roundish objects falling short distances. So for an apple falling from a tree, or a ball dropped from a first floor window, it can be ignored with minor loss of accuracy; but from the fiftieth floor of the Empire State building air resistance may have a significant effect, depending upon how accurately the speed, say, of the falling object is to be calculated! Sky divers rely on reaching a terminal velocity soon after leaving an aircraft. They then fall at a constant speed before opening their parachutes. At the terminal velocity the force of gravity downwards is just balanced by the upward resistance forces of the air. Parachutes ensure that the air resistance equals the parachutist's weight at low speeds.

Let us suppose, however, that for the calculations of this section and the next air resistance can be ignored.

In this section I shall calculate the motion of an apple, say, falling from a tree. I will work out the time it takes to fall to the ground from a height of 3 m and the velocity it reaches by the time it hits the ground.

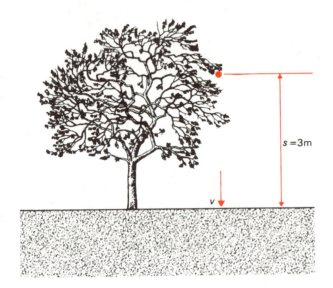

$s = 3\text{m}$

$v$

*Figure 15   A falling apple.*

The first thing I must do is decide upon the direction, either up or down, which I am going to regard as positive. The equations in Section 3.2 presuppose that all the variables have the same positive direction. Because the apple falls downwards I shall regard the downwards direction as positive. Downwards velocities, accelerations and distances from the initial position on the tree are all to be regarded as positive. (I could equally choose upwards as positive, and would obtain the same results, but I must use whatever convention I choose consistently.)

We know the acceleration downwards is $g$, the acceleration due to gravity ($9.81 \text{ m s}^{-2}$). To calculate the time taken for the apple to fall I must choose the best equation from the set of four given earlier. We know the initial velocity $u_y$ is zero, the acceleration is $g$, the distance $s$ is 3 m and we want to calculate $t$. The appropriate equation is equation 13.

$$s = ut + \tfrac{1}{2}at^2 \qquad (t \geqslant 0)$$

Substituting the correct values gives, using SI units:

$$3 = 0 + \tfrac{1}{2} \times 9.81 T^2$$

So $\qquad T^2 = \dfrac{3 \times 2}{9.81}$

and $\qquad T = \pm 0.78$

Because we are only concerned with times that are positive

$$T = 0.78$$

So the apple takes 0.78 s to fall.

The final velocity $v_y$ before hitting the ground is the acceleration multiplied by the time

$$v_y = gt$$

so $V_y = 9.81 \times 0.78 = 7.7$ (to two significant figures) and the velocity on hitting the ground is $7.7 \text{ m s}^{-1}$.

26

In fact, the apple might take slightly longer and the velocity would be a little less owing to the presence of a small amount of air resistance.

*Vertical motion: projecting a ball upwards*

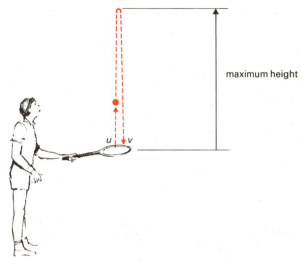

maximum height

*Figure 16    A ball being projected vertically upwards.*

If you throw a ball straight up into the air, how far up will it go, how long will it take to reach its maximum height, with what velocity will it return and after how long? I am going to try to answer all these questions both in general and for particular cases.

(a)  Maximum height

Because the ball is travelling upwards to begin with I shall choose to regard everything *upwards* as positive this time. So the initial velocity is $u_y$ and the acceleration $a = -g$ ($g$ acts downwards not upwards). At maximum height the final velocity $v_y$ must be zero (it is not moving up or down) and I want to know the distance to the top of the flight. I want an equation containing $u, v, a$ and $s$. Equation 14 is the one

$$v^2 = u^2 + 2as$$

If I substitute $v = 0$ and $a = -g$, $u = u_y$ and $s = s_y$, I get

$$0 = u_y^2 - 2gs_y$$

adding $2gs_y$ to both sides gives

$$2gs_y = u_y^2$$

So      $s_y = u_y^2/2g$

The ball will rise to a height $\dfrac{u_y^2}{2g}$. For any particular throw a value for $u_y$ can be substituted into the equation and so can the value for $g$ ($9.81\,\mathrm{m\,s^{-2}}$).

**SAQ 13**

SAQ 13

If a ball is thrown up with velocity $10\,\mathrm{m\,s^{-1}}$ how high will it go?

(b)  Time to reach maximum height

Again the final velocity $v_y$ is zero, the acceleration $a$ is $-g$, $u_y$ is the initial velocity but $t$ is what is required in this case so I need an equation containing $v, a, u$ and $t$. Equation 11 is the one

$$v - u = at$$

27

Substituting 0 for $v$, $g$ for $a$ and $u_y$ for $u$ gives

$$-u_y = -gt$$

dividing both sides by $-g$ gives

$$\frac{u_y}{g} = t$$

so the ball reaches its maximum height after a time $\frac{u_y}{g}$.

**SAQ 14**

**SAQ 14**

Check that $u_y/g$ has the dimensions of time. Find the time a ball thrown up with initial velocity $10\,\text{m s}^{-1}$ would take to reach its maximum height.

### (c) Velocity of return

When the ball returns it has come back to where it started, its position is the same as its initial position so the distance from its initial position is zero. Its acceleration is $-g$ and its initial velocity is $u_y$. So, because I want to find $v_y$ I want an equation involving $s$, $a$, $u$ and $v$. This is equation 14 again

$$v^2 = u^2 + 2as$$

Substituting $s = 0$ gives me $v_y^2 = u_y^2$. There are two solutions to this equation:

$$v_y = +u_y$$

and $\quad v_y = -u_y$

Both solutions give the velocity at the point where $s = 0$. The first one gives the initial velocity at zero time and is not the one we want. So $v_y = -u_y$ is the solution. The ball returns with equal and opposite velocity.

**SAQ 15**

**SAQ 15**

Find an expression for the time the ball takes to return to its starting point.

### *Horizontal component of motion*

Whereas it is possible for a projectile to move vertically, without any horizontal component (e.g. a ball thrown vertically upwards returns to its starting point) it is not possible for one to move freely through the air horizontally without any vertical component, because gravity is always acting and exerting a force at right angles to horizontal motion. The very words horizontal and vertical mean 'at right angles to' and 'in the direction of' gravity.

The best way to think of the horizontal component of motion of a projectile of some kind is to imagine yourself directly above it looking down from a great height (through a telescope). You will only see the horizontal component. A way to simulate the horizontal component only of a projectile is to slide the projectile on something that exerts very little resistance—like ice—but can exert a vertical force equal and opposite to the gravitational force, for example, a puck in ice hockey or a stone in the game of curling. Ignoring air resistance, the horizontal component of velocity of a projectile is constant. There is no force to either accelerate it or decelerate it, so if $u_x$ is the initial value of this horizontal component (in the $x$-direction) then this is its velocity at all times. The horizontal distance $s_x$ covered in time $t$ is $u_x t$.

$$s_x = u_x t$$

*The motion of a projectile expressed as the vector sum of horizontal and vertical motion*

We now have equations describing both the vertical and horizontal components of motion of a projectile. The equations for velocity and distance are

vertically          horizontally

$$v_y = u_y - gt \left.\vphantom{\begin{array}{c}a\\b\end{array}}\right\} \quad (20) \qquad\qquad u_x \text{ is constant} \left.\vphantom{\begin{array}{c}a\\b\end{array}}\right\} \qquad (21)$$

and $\qquad s_y = u_y t - \tfrac{1}{2}gt^2 \qquad\qquad\qquad\qquad s_x = u_x t$

Thus, at any time $t$ the *distance* co-ordinates of the object will be $(u_x t, u_y t - \tfrac{1}{2}gt^2)$ from equations 20 and 21.

What sort of path does the projectile follow? I will put some values in. Suppose $u_x = 10\,\mathrm{m\,s^{-1}}$, $u_y = 20\,\mathrm{m\,s^{-1}}$ and $g = 9.8\,\mathrm{m\,s^{-1}}$. Here is a table of $s_x$ and $s_y$ for different values of $t$.

| $t$/s | 0 | 1 | 2 | 3 | 4 | 5 | $T$ |
|---|---|---|---|---|---|---|---|
| $s_x$/m | 0 | 10 | 20 | 30 | 40 | 50 | $10T$ |
| $s_y$/m | 0 | 15.1 | 20.4 | 15.9 | 1.5 | $-22.5$ | $20T - 4.9T^2$ |

These values are plotted on a graph in Figure 17. The projectile follows a parabolic path. This is true for any values of $u_x$ and $u_y$ (except zero).

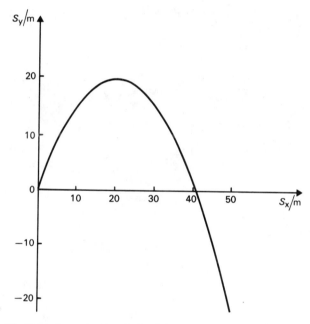

*Figure 17  The flight of a projectile calculated from the equations of motion under constant vertical acceleration.*

What is the (net) initial velocity $u$ of the projectile in this case?

$$u = \sqrt{u_x^2 + u_y^2}$$

$$= \sqrt{100 + 400}$$

$$= \sqrt{500}$$

$$\simeq 22.4\,\mathrm{m\,s^{-1}}$$

It is not necessary to draw a graph to find the distance a projectile travels. It can be calculated.

*Example*

A projectile is fired at a velocity of $100\,\mathrm{m\,s^{-1}}$ at an angle of $30°$ to the ground, which is horizontal. How far will it go before hitting the ground again? List the suppositions you make. Use 9.81 m s$^{-2}$ as the value for g, see Figure 18.

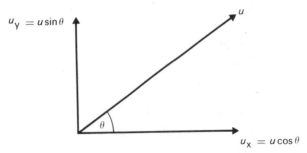

*Figure 18   The horizontal and vertical components of the initial velocity of a projectile.*

The first step is to resolve the motion into vertical and horizontal components.

The vertical component $u_y$ of the initial velocity $u$ is $u \sin \theta$—regarding upwards as the positive direction

So $\qquad u_y = 100 \sin 30° \, \text{m s}^{-1} = 50 \, \text{m s}^{-1}$

Similarly $\qquad u_x = 100 \cos 30° \, \text{m s}^{-1} = 86.6 \, \text{m s}^{-1}$

If I knew the time of flight I could calculate the horizontal distance travelled since $s_x = u_x t$. But first I must calculate the time of flight by considering vertical motion only. Because the projectile returns to the ground the vertical distance of the end of the flight from the start is zero ($s_y = 0$). I know the acceleration is $-g$ (i.e. $-9.81 \, \text{m s}^{-2}$); I know the initial velocity $u$ and want to find $t$. So I will use equation 13; $s = ut + \frac{1}{2}at^2$. Substituting the known quantities in this equation gives

$$0 = 50T - \tfrac{1}{2} \times 9.81 T^2$$

So $\qquad T^2 = \dfrac{50T \times 2}{9.81}$

This can be expressed as

$$T\left(T - \frac{100}{9.81}\right) = 0$$

This equation can be satisfied if either

$$T = 0 \text{ or } T = 100/9.81.$$

That is,

$$T = 0 \quad \text{or} \quad 10.2 \text{ (to three significant figures)}$$

So the time of flight is 10.2 s.

I can now use this value to find the horizontal distance $s_x$.

$\qquad s_x = u_x t$

But $\qquad u_x = u \cos 30° = 86.6 \, \text{m s}^{-1}$

So $\qquad S_x = 86.6 \times 10.2$

$\qquad\qquad = 883$ (to three significant figures)

So this projectile will fly a horizontal distance of 883 m before hitting the ground supposing that (a) air resistance is negligible (b) the variation of g with height is negligible, and (c) any effects of spin are negligible.

### SAQ 16

Another projectile is fired at a velocity of 100 m s$^{-1}$ at an angle with the horizontal of 45°. Calculate, using the same suppositions, how far this one will travel. What height will the projectile reach?

*The equation describing projectile flight*

The equation describing projectile flight can be obtained by eliminating $t$ from the equations for distance; equations 20 and 21.

$$s_y = u_y t - \tfrac{1}{2}gt^2 \tag{20}$$

$$s_x = u_x t \tag{21}$$

provided $u_x \neq 0$ (i.e. the projectile does not go straight up) I can divide equation 21 by $u_x$ giving

$$\frac{s_x}{u_x} = t$$

substituting for $t$ in equation 20 gives

$$s_y = u_y\left(\frac{s_x}{u_x}\right) - \frac{g}{2}\left(\frac{s_x}{u_x}\right)^2 \tag{22}$$

This is a quadratic equation. It is the equation of a parabola. You are shown how to solve it for $s_x$, given all the other quantities, in Section 4.1.

### SAQ 17

Use equation 22 to check the values for $s_y$ obtained in the worked example and in SAQ 16.

*The angle at which to throw a ball in order to obtain maximum range*

You should have found that a 45° angle of projection gives a greater distance than 30°. Would 60° be even better? If a ball is thrown from about ground level with an initial velocity $u$, what is the best angle at which to throw it in order to make it go as far as possible? I want to explain how you can answer this question without just using trial and error.

If the speed is $u$ and the projectile is thrown at an angle $\theta$ to the horizontal ($0 < \theta < 90°$) then $u_x = u\cos\theta$ and $u_y = u\sin\theta$ (Figure 18) as before.

Again I can calculate the time of flight by working out when the projectile will hit the ground. By considering vertical motion only to begin with, I calculate the time at which the vertical distance travelled by the projectile is zero again.

that is    $s_y = 0$

I know that the initial velocity is $u_y$ and the acceleration is $-g$

So        $0 = u_y t - \tfrac{1}{2}gt^2$ (using $s = ut + \tfrac{1}{2}at^2$)

i.e.      $0 = t(u_y - \tfrac{1}{2}gt)$

This equation is solved (i.e. the right-hand side of the equation is zero) if $t = 0$ or if $(u_y - \tfrac{1}{2}gt)$ is zero (i.e. $u_y = \tfrac{1}{2}gt$).

(See Section 4.1 for further discussion of this kind of solution to an equation.)

The result $t = 0$ refers to the start of the flight, so the result I want is

$$u_y = gt/2$$

or        $t = 2u_y/g$

Now I can calculate the horizontal distance. The initial velocity is $u_x$; the time is $2u_y/g$

So        $$s_x = \frac{2u_x u_y}{g}$$

But       $u_x = u\cos\theta$

$$u_y = u\sin\theta$$

So        $$s_x = \frac{2u^2 \sin\theta\cos\theta}{g}$$

31

Now in Unit 4 you were shown that $\sin 2\theta = 2\sin\theta\cos\theta$. I can use this result and write

$$s_x = \frac{u^2 \sin 2\theta}{g}$$

Now I want the angle $\theta$ which gives maximum range (i.e. the largest value of $s_x$ for a given velocity). So I have to find the largest possible value of $\sin 2\theta$.

> What is the largest value of the sine function? Look it up in your sine tables if you are not sure.

Its largest value is 1.

So $$s_{x(max)} = \frac{u^2}{g}$$

The sine of 90° has the value 1, so if $2\theta$ is 90° the value of $\theta$ for maximum range is 45°, see Figure 19.

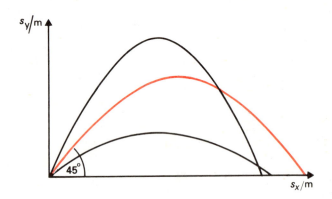

*Figure 19   Illustrating that the angle of projection for maximum is 45° (ignoring air resistance).*

Actually, air resistance tends to affect things quite noticeably. At the speed golf balls travel, an angle much lower than this gives maximum range. With tennis balls and ping pong balls too air resistance is significant and spin affects the flight. Top spin reduces the range of low angle shots, for example. These facts reveal the limitations of the model we have been using. More accurate models including the effects of air resistance can quite easily be devised, as we will see later, but they tend to be much more difficult to solve in a general way. They can be solved by computer however.

### 3.3   Summary

1   The equations of motion along a straight line with constant acceleration are

$$v = u + at \tag{11}$$
$$s = \tfrac{1}{2}(u + v)t \tag{12}$$
$$s = ut + \tfrac{1}{2}at^2 \tag{13}$$
$$v^2 = u^2 + 2as \tag{14}$$
$$(s = vt - \tfrac{1}{2}at^2)$$

2   These equations are particularly applicable to the motion of projectiles since a fairly accurate model of vertical motion under the influence of gravitation is one in which acceleration downwards is constant and equal to $g$ (air resistance is neglected). Motion is resolved into vertical and horizontal components.

3   If air resistance can be neglected the angle of projection for a maximum range on horizontal ground is 45° to the horizontal.

# 4 QUADRATIC EQUATIONS AND COMPLEX NUMBERS

A *quadratic equation* is an equation relating two variables which involves the square (but no higher power) of the independent variable. Such equations may be between dimensioned quantities as in the quadratic equation

$$s = ut + \tfrac{1}{2}at^2$$

quadratic equation

or between numbers as in

$$Y = X^2 - 4X + 4.$$

The graph of a quadratic equation is a parabola.

If variables are linked by a quadratic equation it is easy to find the value of the dependent variable given a value of the independent variable. But in some cases, it is necessary to find a value for the independent variable given a value of the dependent variable. For instance, if you were asked to find $X$ when $Y = 1$ in the equation

$$Y = X^2 - 4X + 4$$

then $X$ must satisfy

$$1 = X^2 - 4X + 4$$

which can be written as

$$X^2 - 4X + 3 = 0.$$

This equation is also called a quadratic equation, but is rather different from

$$Y = X^2 + 4X + 3.$$

This latter equation links a whole range of values of $X$ and $Y$ while

$$X^2 - 4X + 3 = 0$$

is only satisfied for two particular values of $X$, the values $X = 1$ and $X = 3$. (Check for yourself by substitution that $X^2 - 4X + 3 = 0$ when $X = 1$ and when $X = 3$.) These values for $X$ which satisfy the quadratic equation are sometimes called the *roots* of the quadratic equation.

This section explains how the roots of quadratic equations may be found. I will show you that such equations usually have two roots, but may have only one root or even no roots. If an equation has no roots then a new kind of number called an imaginary or complex number can be used to solve the equation.

If you think you can solve a quadratic equation and can answer either SAQ 20 or SAQ 21, or both, then you need not read Sections 4.1 to 4.4. If you do not know how to solve a quadratic equation, or did once know and have forgotten, read through these sections and learn either the formula method or the method of completing the square. You don't need both. The formula is included in your Course Handbook.

## 4.1 Graphical solutions

In Section 3.2 you met the equation

$$D_s = 0.6U + \frac{U^2}{13} \qquad (U \geqslant 0)$$

I can use the graph of this function to solve the equation

$$100 = 0.6U + \frac{2U^2}{13} \qquad (U \geqslant 0)$$

by reading off the value of $U$ for which $D_s = 100$.

This method is time consuming and not very accurate, but it illustrates why a quadratic equation may have two solutions.

**SAQ 18**

Plot a graph of $Y = X^2 + 4X$ on graph paper for values of $X$ from $-5$ to 3. Read off the solutions to the following equations. (*Note:* there may be more than one solution for each case.)

(a) $X^2 + 4X = 0$

(b) $X^2 + 4X - 3 = 0$

(c) $X^2 + 4X + 4 = 0$

(d) $X^2 + 4X + 5 = 0$

You will notice that there may be no solution, one solution or two solutions to a quadratic equation. There is no solution if the parabola does not reach the value of $Y$; there is one if it just reaches it, and there are two if the parabola passes through the value twice. When the parabola just touches the value of $Y$ and there is one solution this one solution is sometimes called, rather confusingly, a *double solution*. By this is meant that the two solutions are coincident—occur at one value of $X$.

double solution

## 4.2  Completing the square

We could have solved some of these equations more quickly without using the graph. The case of $X^2 + 4X = 0$ can be solved easily because $X$ is a factor in each term. The equation can be written as

$X(X + 4) = 0$

Now the whole expression will have the value zero if *either* $X$ *or* $(X + 4)$ is zero. So there are two solutions

Either   $X = 0$

or   $X + 4 = 0$ so $X = -4$

So the two solutions, or roots, are $X = 0$ or $-4$.

The values of $X$ for which the other equations in SAQ 18 are true are not so easily obtained. One method, called 'completing the square', involves rewriting the equation so that the terms in $X^2$ and $X$ form part of a squared term. For example, $X^2 + 4X$ is written as

$(X + 2)^2 - 4$

because $(X + 2)^2 = X^2 + 4X + 4$ and so $(X + 2)^2 - 4 = X^2 + 4X$.

So referring to SAQ 18 again

(a) becomes $(X + 2)^2 - 4 = 0$
   or   $(X + 2)^2 = 4$

(b) becomes $(X + 2)^2 - 4 - 3 = 0$
   or   $(X + 2)^2 = 7$

(c) becomes $(X + 2)^2 - 4 + 4 = 0$
   or   $(X + 2)^2 = ?$

(d) becomes $(X + 2)^2 = ?$

What are the numbers on the right-hand side of the new equations for (c) and (d)?

(For the correct answers see the text below under (c) and (d).)

Having achieved this step, of 'completing the square', the next step is simply to take the square root of both sides of the equation. For example, in the case of (a)

$$(X + 2)^2 = 4$$

So $\quad X + 2 = \pm\sqrt{4} = \pm 2$

Remember that $(-2)^2 = 4$ as well as $(+2)^2$ so you have to write $\pm$ in front of the number 2 to indicate that the square root of 4 is either $+2$ or $-2$.

But if $\quad X + 2 = \pm 2$

$$X = (-2 + 2) \quad \text{or} \quad (-2 - 2)$$

So $\qquad X = 0 \quad \text{or} \quad -4$

### SAQ 19 SAQ 19

What are the solutions for the case of equation (b) $(X + 2)^2 = 7$?

Equation (c) becomes $(X + 2)^2 = 0$. On taking the square root you obtain

$$X + 2 = 0$$

so $\qquad X = -2$

This is a so-called 'double solution' because both values of $X$ are the same.

Equation (d) becomes $(X + 2)^2 = -1$

Now if you take the square root you obtain

$$X + 2 = \pm\sqrt{-1}$$

There is no normal number which, when squared, will have the value $-1$. This then is an example of an equation with no normal solution. However, in Section 4.5 I will briefly explain how 'complex numbers' have been invented to give solutions to such equations.

### SAQ 20 SAQ 20

Try to complete the square in the following examples and hence solve each equation, or state there is no solution which satisfies the equation.

(a) $\quad X^2 + 2X + 1 = 0$

(b) $\quad X^2 + 2X = 0$

(c) $\quad X^2 - 2X - 3 = 0$

(d) $\quad 3X^2 - 12X - 1 = 0$

(Use your slide rule to calculate the square root in part (d) to three significant figures.)

### 4.3 The Formula method

If we complete the square for the general quadratic equation $0 = ax^2 + bx + c$, where $a$, $b$ and $c$ are any parameters, $a$ having a non-zero value, then we can derive a formula for obtaining the solutions to any quadratic equation. You do not need to be able to reproduce the following argument, but you should follow it through to see where the formula comes from.

First divide each term of the equation by $a$ so that $x^2$ has a coefficient of 1.

So $\qquad ax^2 + bx + c = 0$

becomes $\quad x^2 + \dfrac{b}{a}x + \dfrac{c}{a} = 0$

Now to complete the square you first pay attention to the terms in $x^2$ and $x$.

$$x^2 + \frac{b}{a}x \quad \text{can be written as}$$

$$\left(x + \frac{b}{2a}\right)^2 - \frac{b^2}{4a^2}$$

Now, adding in the last term again you can reconstruct the original equation, but in the form of the 'completed square'. Thus

$$\left(x + \frac{b}{2a}\right)^2 - \frac{b^2}{4a^2} + \frac{c}{a} = 0$$

Re-arrange, leaving the squared term on its own on the left-hand side

$$\left(x + \frac{b}{2a}\right)^2 = \frac{b^2}{4a^2} - \frac{c}{a}$$

or $\quad \left(x + \frac{b}{2a}\right)^2 = \frac{b^2 - 4ac}{4a^2}$

Take the square root of each side

$$x + \frac{b}{2a} = \pm\sqrt{\frac{b^2 - 4ac}{4a^2}}$$

The square root of $4a^2$ is $2a$

So $\quad x = \frac{-b}{2a} \pm \frac{\sqrt{b^2 - 4ac}}{2a}$

or, in its usually quoted form, with a common denominator of $2a$

$$x = \frac{-b \pm \sqrt{b^2 - 4ac}}{2a} \tag{23}$$

If $b^2 > 4ac$ then there will be two distinct solutions

If $b^2 = 4ac$ there will be a 'double solution'

If $b^2 < 4ac$ there will be no real solution because the number under the square root sign is negative.

If the quadratic equation is quoted in terms of numbers, $AX^2 + BX + C = 0$, then equation 23 becomes $X = \dfrac{-B \pm \sqrt{B^2 - 4AC}}{2A}$.

*Example*

Solve the equation $2X^2 - 4X + 1 = 0$

For this equation to be the same as $AX^2 + BX + C = 0$, $A = 2$, $B = -4$ and $C = 1$. We can now substitute these values in the general equation

giving $\quad X = \dfrac{-(-4) \pm \sqrt{(-4)^2 - 4 \times 2 \times 1}}{2 \times 2}$

So $\quad X = \dfrac{4 \pm \sqrt{16 - 8}}{4} = 1 \pm \dfrac{\sqrt{8}}{4} = 1 \pm 0.707$

So $\quad X = 1.71$ or $0.293$ (to three significant figures)

**SAQ 21**

Use the formula method to solve

(a) $X^2 + 6X = 0$

(b) $9X^2 + 12X + 2 = 0$

(c) $3X^2 - 6X + 10 = 0$

## 4.4 Factorization

Sometimes a quadratic equation can be solved easily by a process of factorization. Consider the first equation in SAQ 21. $X$ appears in both of the terms on the left-hand side of the equation and is said to be a common factor. Therefore, the equation $X^2 + 6X = 0$ can be written as

$$X(X + 6) = 0$$

This process of expressing a number of terms as the product of their factors is called factorization.

The equation is satisfied if either $X = 0$ or $X + 6 = 0$. So the solution is

$$X = 0 \quad \text{or} \quad -6$$

Some equations can be written as the product of two simple factors, for example, $(3X - 2)(X + 1) = 0$. This equation is satisfied if $X + 1 = 0$ or if $3X - 2 = 0$. If either bracket is zero the equation is solved.

So $\quad X = -1 \quad$ or $\quad 2/3$.

But $(3X - 2)(X + 1)$ are the factors of $3X^2 + X - 2$ so the equation $3X^2 + X - 2 = 0$ has the solutions $X = -1$ or $2/3$.

If you are lucky the equation you have to solve may have convenient factors of this kind, in which case a solution is readily found. Most equations are less helpful than this in practice!

The following equations have simple factors. Try to find them and hence solve the equations. (This is not set as an SAQ because this method of solution is not one of the objectives of the course.)

    (a) $\quad X^2 + 5X + 6 = 0$

    (b) $\quad X^2 - X - 6 = 0$

    (c) $\quad 2X^2 + 5X - 3 = 0$

If you cannot factorize them use the formula to solve the equations. Check your results by writing the factors and multiplying them out.

## 4.5 Imaginary and complex numbers

As you have seen, some equations do not seem to have solutions. For example, when completing the square of equation $X^2 + 4X + 5 = 0$ you obtain $(X + 2)^2 = -1$, but you can proceed no further because you have no way of taking the square root of $-1$. If you use the formula to get the solution you will find that

$$X = \frac{-4 \pm \sqrt{16 - 20}}{2} = \frac{-4 \pm \sqrt{-4}}{2}$$

but again you cannot obtain the square root of $-4$.

We give meaning to symbols like $\sqrt{-1}$ and $\sqrt{-4}$ by inventing a new system of *imaginary numbers*. By using these imaginary numbers mathematical manipulation can sometimes be simplified and 'imaginary' solutions to equations can be used to model real physical situations. A whole new topic has now grown up in mathematics which is concerned with imaginary numbers. Ordinary numbers of the kind we have been using hitherto are called *real* numbers to distinguish them.

*imaginary numbers*

So let us suppose that $\sqrt{-1}$ is a new kind of number, which is not a real number, whether positive, negative or zero, and call it an imaginary number i.* It is defined by the fact that

$$i \times i = -1$$

---

*In engineering the square root of $-1$ is usually called j instead of i. Again, by definition, $j \times j = -1$

Thus, $i \times i \times i = -1 \times i = -i$ and $i \times i \times i \times i = (-1) \times (-1) = 1$.

Imaginary numbers can be added together and can be multiplied by real numbers so that

$$i + i + i + i = 4i$$

Combining these two rules gives, for example, $2i \times 2i = -4$. Indeed, the quantity $i$ is treated as if it were a new real number, except that when multiplied by another imaginary number it produces a *real* number and that when added to a real number, or subtracted from it there is no further reduction possible.

If you add a real number to another real number, say $2 + 3$, you can reduce this expression to the number 5. But if you add an imaginary number to a real number no further reduction is possible. $2 + i$ cannot be further reduced and is called a *complex number* with a real part and an imaginary part. complex number

We can now solve the equations which previously we referred to as having no solutions, by finding their *imaginary* or *complex* solutions

Thus, if $(2 + X)^2 = -1$

$$2 + X = \pm\sqrt{-1}$$

That is $2 + X = \pm i$

So there are two solutions

either $X = -2 + i$

or $X = -2 - i$

Each solution is a complex number, $-2$ is the real part and $+i$ or $-i$ is the imaginary part.

Again when $X = \dfrac{-4 \pm \sqrt{-4}}{2}$

we can write the square root sign as a product of a real and imaginary number. Thus, $\sqrt{-4}$ can be written as $\sqrt{4} \times \sqrt{-1}$.

So $X = \dfrac{-4 \pm \sqrt{4}\sqrt{-1}}{2}$

$= \dfrac{-4 \pm 2i}{2}$

$= -2 \pm i$

which is the same solution as before.

**SAQ 22** SAQ 22

Find the solutions of the following equations

(a) $X^2 + 2X + 2 = 0$   (b) $X^2 - 4X + 7 = 0$   (c) $X^2 + 5X + 3 = 0$

This section is simply a brief introduction to the idea of complex and imaginary numbers. You will return to the topic later in the course. You might, however, find the following representation of imaginary numbers helpful. It suggests that a complex number can be thought of as a vector.

Suppose you regard the straight line shown in Figure 20 as representing all the possible real numbers that exist. It tends to infinity in either direction, and no real numbers lie anywhere other than on this line. Positive real numbers are to the right of the point representing zero, negative ones are to

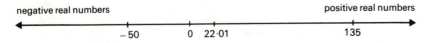
negative real numbers · positive real numbers · −50 · 0 · 22·01 · 135

*Figure 20   A straight line representation of all real numbers, both positive and negative.*

the left. It is a kind of one-dimensional world: any excursion from it can only be done in the imagination, like 'time travelling' in science fiction.

We can now add an imaginary dimension to this diagram, as in Figure 21, in which the new line represents all the 'possible' *imaginary* numbers. 2i, 5000i, 4.321i are all above zero; $-3i$, $-3.26i$, etc. are all below it.

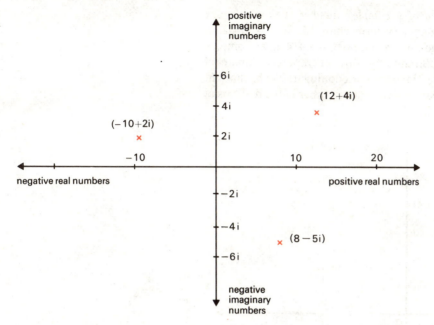

*Figure 21   An Argand diagram, showing the real and imaginary axes and the representation of complex numbers by points on the diagram.*

This diagram is called an *Argand diagram*, after its originator, and complex numbers are represented by points in the two-dimensional plane defined by these axes. The diagram shows a few.

Argand diagram

**SAQ 23**

SAQ 23

Represent the complex numbers

$1 + i, 3 - i, -2 + 3i$ on the diagram of Figure 21.

In connection with this Argand diagram two further important ideas should be introduced at this stage for use in mathematical models later in the course.

First, the magnitude of a complex number, which is usually called the *modulus*, is the length of the line from the point representing it on the Argand diagram to the origin. This is just like the magnitude of a vector. As shown in Figure 22, the modulus of $4 + 3i$ is $\sqrt{4^2 + 3^2} = 5$. (The square of the length

modulus

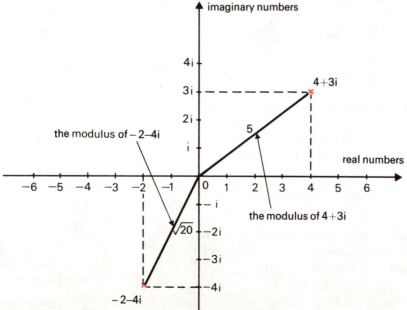

*Figure 22   An Argand diagram showing the magnitudes of the complex numbers $4 + 3i$ and $-2 - 4i$.*

of the hypotenuse is equal to the sum of the squares of the other two sides— Pythagoras.) Similarly, the modulus of the complex number $-2 - 4i$ is $\sqrt{2^2 + 4^2} = \sqrt{20} = 4.47$, to three significant figures. So, generally the modulus of $A + Bi$ is $\sqrt{A^2 + B^2}$. The modulus of the complex number $A + Bi$ is often written $|A + Bi|$.

Secondly, the *complex conjugate of a complex number* is that complex number with the sign of the imaginary term changed. So $3 + 2i$ has a complex conjugate $3 - 2i$, and the complex number $-5 + 4i$ has a complex conjugate $-5 - 4i$. Indeed $(A + Bi)$ and $(A - Bi)$ are complex conjugates of each other. As indicated in Figure 23 the complex conjugate can be thought of as the reflection of the corresponding complex number in the real axis of the Argand diagram.

complex conjugate of a complex number

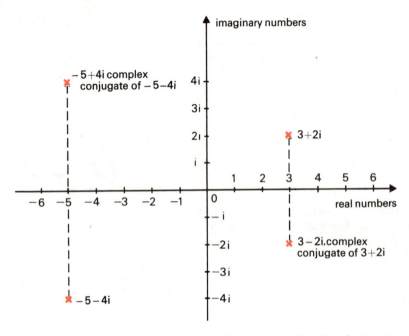

Figure 23   An Argand diagram showing the complex conjugates of $3 + 2i$ and $-5 - 4i$.

### SAQ 24

SAQ 24

What are (a) the moduli and (b) the complex conjugates of the following complex numbers:

      (i)    $5 + 12i$

      (ii)   $2 - 3i$

      (iii)  $-A - Bi$

What are the moduli of the complex conjugates?

### 4.6   Maximum and minimum values of parabolas

Figures 24(a) and (b) illustrate what is meant by the maximum or minimum values of parabolas. The maximum value is the value of $Y$ at the top of the hump in a parabola which opens downwards like the one in Figure 24(a); the minimum is the value of $Y$ at the bottom of the hollow in a parabola which opens upwards like the one in Figure 24(b).

Suppose you were asked to find the minimum value of the parabola

    $Y = X^2 - 2X - 3.$

You could do it by plotting the graph of the parabola, as in Figure 25. But there is also an algebraic way of doing it. Suppose at the minimum point $Y$ has the value $Y_1$ and $X$ has the value $X_1$. Then

    $Y_1 = X_1^2 - 2X_1 - 3.$

40

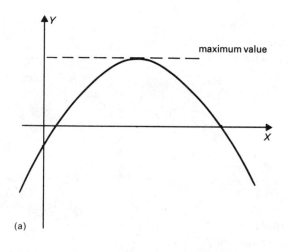

Figure 24   The maximum and minimum values of parabolas.

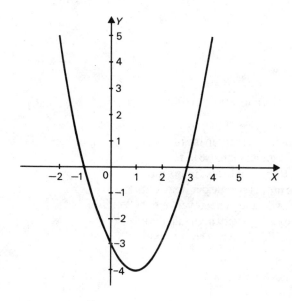

Figure 25   The parabola $Y = X^2 - 2X - 3$.

You have already seen in Section 4.1 that there is a so-called 'double solution' when the parabola just touches a value of $Y$. When $X = X_1$ the parabola does just touch $Y_1$ and so there must be a double solution for $X_1$ in the quadratic equation

$$Y_1 = X_1^2 - 2X_1 - 3.$$

You might recognize it more clearly as a quadratic if I write it as

$$0 = X_1^2 - 2X_1 - 3 - Y_1.$$

You know from Section 4.3 when a 'double solution' occurs—it occurs for the quadratic

$$AX^2 + BX + C = 0$$

when $B^2 = 4AC$, and in that case $X = -B/2A$.

In this case, then, there is a 'double solution' for $X_1$ when

$$X_1 = \frac{-(-2)}{2 \times 1} = 1.$$

If $X_1 = 1$ then $Y_1 = -4$, so the minimum value is at $Y = -4$, when $X = 1$.

This is a general algebraic method of finding the maximum or minimum value of any parabola. If the parabola has the equation

$$Y = AX^2 + BX + C$$

then the maximum or minimum will occur at $X = -B/2A$.

41

A quick sketch of the parabola in question will indicate whether it has a maximum or a minimum point. Alternatively, as you have seen in Unit 5, if $A$ is *positive* then the parabola opens upwards and has a *minimum*, while if $A$ is *negative* the parabola opens downwards and has a *maximum* value.

**SAQ 25**

SAQ 25

Will the parabola

$$Y = 1 + 4X - 3X^2$$

have a maximum value or a minimum value? Find the value of $Y$ and the value of $X$ at which one or the other occurs.

You will come across other ways of finding maximum or minimum values of mathematical expressions in Unit 7. The method given in this section only works for parabolas.

### 4.7 The level of tax giving the maximum revenue

This section presents an illustration of the use of quadratic equations.

An increase in the tax on a certain item puts up the price. If the tax is small compared with the price then there will not be much difference in the number of these items sold, so an increase in tax should mean an increase in revenue for the government. If, however, the tax is large, then an increase in tax may not actually result in more revenue. There will be more revenue per item, but there may be less demand for the item and so fewer sold. The government might in fact get less total return in the end by imposing a very high tax, than by imposing a small tax. The model used in Unit 3 leads to a description of tax yield in terms of a quadratic function.

The model in Unit 3 uses a linear model for the supply

$$q_s = a + b(p_s - t) \qquad (24)$$

$q_s$ being the quantity of items that producers are willing to supply to sell at price $p_s$. (The producer must produce the item at a pre-tax price of $p_s - t$.) $a$ and $b$ are parameters, both greater than zero. The demand is modelled by

$$q_d = c - dp_d \qquad (25)$$

$q_d$ being the quantity consumers would buy at a selling price of $p_d$ ($c > 0, d > 0$). The equilibrium price $\bar{p}$ and equilibrium quantity $\bar{q}$ should satisfy both equations. That is, there is a price $\bar{p}$ and a quantity $\bar{q}$ which, when substituted for $p_s$, $p_d$ and for $q_s$, $q_d$, respectively, will satisfy both equations simultaneously.

That is $\bar{q} = a + b(\bar{p} - t)$

$\bar{q} = c - d\bar{p}$

**SAQ 26**

SAQ 26

Solve these two simultaneous equations in $\bar{q}$ and $\bar{p}$ for the value of $\bar{p}$ in terms of $a$, $b$, $c$, $d$ and $t$.

The solution of these equations for $\bar{q}$ is

$$\bar{q} = \frac{bc + da - bdt}{b + d}$$

Now, the tax collected per item is $t$, and since $\bar{q}$ items will be bought the revenue $r$, from tax on them will be $\bar{q}t$

So
$$r = \left(\frac{bc + da - bdt}{b + d}\right)t$$

$$= \frac{(bc + da)}{b + d}t - \frac{bd}{b + d}t^2$$

This is a quadratic equation in $t$.

What value should $t$, the tax per item, have to make the revenue a maximum?

Because the term in $t^2$ has a negative sign the graph of this equation is a parabola with a maximum value. I want to find the value of $t$ at the maximum, preferably without drawing an accurate graph. As you saw earlier, one way to do this is to find the value of $t$ which gives us a double solution, because this occurs at the maximum.

The general formula for the 'double solution' of the parabola with an equation '$a$'$t^2$ + '$b$'$t$ + '$c$' = $r$ is

$$t = -\frac{'b'}{2'a'}$$

where I have put inverted commas round '$a$', '$b$' and '$c$' to avoid confusion with the parameters $a, b$ and $c$ in this taxation model.

Now '$a$' is the coefficient of $t^2$, that is, the number multiplying $t^2$. So
'$a$' $= \dfrac{-bd}{b + d}$.

Similarly, '$b$' is the coefficient of $t$, so '$b$' $= \dfrac{bc + da}{b + d}$.

The revenue will be a maximum, then when

$$t = \left(\frac{bc + da}{b + d}\right) \div \frac{2bd}{b + d}$$

$$t = \frac{bc + da}{2bd}. \tag{26}$$

So this is the best value of tax to charge in order to maximize the revenue.

### SAQ 27

SAQ 27

Suppose a government department were trying to work out what tax to recommend on petrol. Market research has shown that the quantity $Q$ gallons of petrol customers would be willing to buy at a price $P$ pence per gallon ($0 < P < 200$) was well approximated by the linear equation

$$Q = 400\,000 - 1500\,P.$$

A survey of oil companies showed that the quantity $Q$ gallons of petrol which they were willing to produce to sell at a price $P$ pence per gallon with tax $T$ pence per gallon was well approximated by the linear equation

$$Q = 100\,000 + 8300(P - T) \qquad (10 < P < 200).$$

On this evidence what tax rate should be recommended to the government to produce the maximum tax yield?

### 4.8  Summary

1  There are four methods of solving quadratic equations

(a)  graphically

(b)  by completing the square. The terms in $x^2$ and $x$ are expressed as a squared term. Thus, $x^2 + bx$ is written as $\left(x + \dfrac{b}{2}\right)^2 - \dfrac{b^2}{4}$.

43

(c) by using the formula $x = \dfrac{-b \pm \sqrt{b^2 - 4ac}}{2a}$ (23)

(d) by factorization

(You should be able to use either the formula method or the method of completing the square.)

2 The square root of $-1$ is called the imaginary number i.

$$i \times i = -1$$
$$i \times i \times i = -i$$
$$i \times i \times i \times i = 1$$
$$i + i + i = 3i$$
$$\sqrt{-4} = \sqrt{4} \times \sqrt{-1} = 2i$$

3 A complex number is the sum of a real number and an imaginary number $A + Bi$. Quadratic equations with no real roots have complex roots (i.e. roots which are complex numbers). The magnitude of $A + Bi$ is $\sqrt{A^2 + B^2}$. The complex conjugate of $A + Bi$ is $A - Bi$.

4 The value of $x$ in the quadratic equation $ax^2 + bx + c = y$ which gives the maximum or minimum value of $y$ is $x = \dfrac{-b}{2a}$.

(If $a$ is a negative number the value of $y$ reaches a maximum. If $a$ is a positive number the value of $y$ reaches a minimum.)

# 5  TRAFFIC AND SAFETY

Earlier we derived a general equation for the stopping distance to be expected for any given initial velocity. In this section I want to extend this study to embrace certain other aspects of safety and to show how simple models can lead to simple calculations which, in turn, can be of value in ensuring safety in car travel. This will involve the solution of quadratic equations.

## 5.1  Skid marks

When the police are trying to reconstruct an accident, the skid marks of a vehicle can be very informative. From the length of the skid marks of the wheels, the police can estimate the speed at which the vehicle was travelling before the wheels locked and the vehicle went into the skid.

To evaluate the road surface the police drive a similar 'test' car with similar tyres and cause it to skid at the same place, but at a lower speed. They then compare the skid marks they produce with the original ones. They assume that the frictional forces between surfaces are not dependent upon the speed of the car, only upon the mass of the car, the condition of the road and the type of surface. The frictional force is assumed to be proportional to the mass of the vehicle, as are any accelerating or decelerating forces due to gravity if the car is going up or down hill. The retardation of both cars is the same according to these suppositions.

For the test car, the initial velocity $u_t$, the final velocity, $v_t = 0$, and the distance of skid $s_t$ are known. How can the acceleration be found?

Use $\quad v^2 = u^2 + 2as$

So $\quad 0 = u_t^2 + 2as_t$

So $\quad a = \dfrac{-(u_t)^2}{2s_t}$ $\hspace{4cm}$ (27)

The deceleration of the original car is assumed to be the same; its final velocity is also assumed to be zero; the distance $s$ that it skidded is known, so what is its initial velocity?

Use $v^2 = u^2 + 2as$ again. But we know what $a$ is from equation 27.

So $\quad 0 = u^2 - \dfrac{2u_t^2}{2s_t}s$

i.e. $u^2 = u_t^2\,\dfrac{s}{s_t}$ or $u = u_t\sqrt{\dfrac{s}{s_t}}$

So the speed of the car before the accident can be estimated. Note that the driver's original speed would be greater than this if the car did not stop at the end of the skid.

The ratio of the two velocities can be written as

$$\frac{u}{u_t} = \sqrt{\frac{s}{s_t}}$$

The ratio of the velocities is the square root of the ratios of the skid distances.

After an accident in a built-up area, the skid marks of the car involved were of length 72 m. The test car, initially travelling at $10\,\mathrm{m\,s^{-1}}$, skids to a halt in a distance of 20 m on the same stretch of road. Estimate the initial speed of the car in the accident. Comment on the suppositions made and on the reliability of your estimate.

## 5.2 Traffic flow

Models of traffic flow help to show how to avoid hold ups and to ensure optimum flow in congested conditions. When a queue of vehicles is moving along a road, an increase in its speed usually results in an increase in the distance between the vehicles, so it does not follow that greater speed necessarily means a greater flow of traffic. The distance a driver maintains between his own car and the car in front depends upon his estimate of what is safe. The safest minimum distance for cars to be apart is the stopping distance for the particular speed and road conditions at which they are travelling. In practice, *The Highway Code* recommends at least one car's length for every 10 mph, which is just about the thinking distance (Section 3.2). The reasoning behind this is that after a car's brake lights go on it will not come to a halt until it has travelled the braking distance. If the car behind sees the brake lights it should be able to stop within the *stopping* distance (the braking distance plus the thinking distance). So if the cars are the thinking distance apart each car should just be able to pull up in time to avoid hitting the one in front. In practice, since the brakes of different cars vary in efficiency the minimum distance between cars should be a bit greater than this. This suggests trying two models of traffic flow, one at each extreme: one which assumes the cars are on average separated by the stopping distance and one which assumes they are on average separated by the thinking distance. In queues of slow moving vehicles, however, or in fog, the separation is possibly more or less constant. So several different patterns of behaviour are possible.

I shall consider flow rates based on three different models:

(a) separation is independent of speed, i.e. the distance between cars is constant;

(b) separation is the thinking distance;

(c) separation is the stopping distance.

I shall try to find out if there are speeds which give maximum flow rates, and thus perhaps discover what kinds of roads are appropriate for congested conditions.

These models all have several suppositions in common. They suppose that all the vehicles in the queue are moving with the same constant velocity $v$, and that the distance between the rear bumpers of successive vehicles is the same distance $d$. They also assume there is no possibility of cars overtaking, turning off or parking.

First I must obtain an expression for traffic flow for each model. When one car has just passed a given point the next car is a distance $d$ behind. Travelling with a velocity $v$ it will take this car a time $d/v$ to cover this distance. So after a time $d/v$ it too will have just passed the point. So the cars pass at intervals of $d/v$ (Figure 26). Now if cars pass at intervals of half a second, two cars pass per second; if they pass at intervals of $1/5$ s then 5 pass per second. Similarly, if cars pass at intervals of $d/v$, the flow rate is $v/d$. That is,

$$\text{traffic flow,}\quad f_t = \frac{v}{d} = \frac{\text{velocity}}{\text{distance apart}} \qquad (28)$$

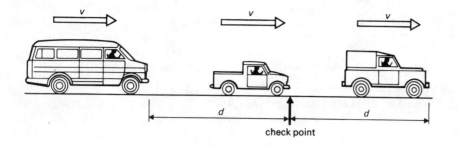

*Figure 26 Traffic moving past a check point. The model supposes that the velocity v of all the cars is the same and that they are spaced at equal distances d. Rate of flow is then $v/d$ cars per second.*

According to the first model the traffic flow is proportional to the velocity, because $d$ is a constant. Thus, using equation 28,

$$f_t = \text{constant} \times v.$$

The faster you go the greater the flow!

According to the second model the distance between the bumpers is the thinking distance $d_t$. But $d_t$ is the velocity $v$ times the thinking time $t_t$ (Section 3.2).

So
$$f_t = \frac{v}{v \times t_t} = \frac{1}{t_t}$$

which means that $f_t$ should be constant. If $t_t = 0.6$ s the flow rate is 1.67 cars per second whatever their speed. The faster drivers go the further apart they are, so the flow should not alter according to this model!

Now this may be approximately true at high speeds, but it's certainly not true at very slow speeds. A slow traffic crawl is also a low flow rate. The reason why the model has failed here is that the distance between cars which determines flow is the distance between, say, the rear bumpers of successive cars, whereas the distance which we call the thinking distance $d_t$ is the distance from the front bumper of one car to the rear bumper of the one in front. A better model then is to suppose that all cars are the same length $b$ so that the distance between rear bumpers is $b + d_t$.

So
$$f_t = \frac{v}{b + d_t} = \frac{v}{b + vt_t}$$

At high speeds $vt_t$ is much bigger than the length of a car $b$, so we can approximately ignore $b$ and $f_t$ is a constant as before. At low speeds $vt_t$ is much less than $b$, so we can ignore $vt_t$ and

$$f_t = \frac{v}{b}$$

Here the flow rate decreases with speed—as it should. The following table is obtained by substituting 0.6 s for $t_t$ and 4 m for $b$:

| $v/\text{m s}^{-1}$ | 0 | 2 | 5 | 10 | 15 | 20 | 30 | 40 | 60 |
|---|---|---|---|---|---|---|---|---|---|
| $f/\text{s}^{-1}$ | 0 | 0.38 | 0.71 | 1.0 | 1.15 | 1.25 | 1.36 | 1.43 | 1.5 |

Notice the increase in traffic flow between $5\,\text{m s}^{-1}$ and $15\,\text{m s}^{-1}$ is $0.44\,\text{s}^{-1}$ whereas between $30\,\text{m s}^{-1}$ and $40\,\text{m s}^{-1}$ it is only $0.07\,\text{s}^{-1}$! Hence, according to this model an attempt to increase flow to above about 1.5 cars per second by increasing speed would not succeed; more traffic lanes would be needed.

According to the third model (c) the distance between successive cars $d$ is $b + d_s$, the car length plus the stopping distance.

Now we found earlier that the stopping distance is given by

$$d_s = \frac{v^2}{2a} + vt_t \qquad \text{(from equation 18)}$$

where $a$ is the maximum deceleration which can be achieved by braking—about $6.5\,\text{ms}^{-2}$. So, according to this model,

$$f_t = \frac{v}{b + vt_t + v^2/2a} \qquad (29)$$

When $vt_t + v^2/2a$ is much smaller than $b$

$$f_t = \frac{v}{b}$$

So the flow increases with speed as in the previous model. As $v$ increases the dominant term of the denominator becomes $v^2/2a$ and hence, at high speeds

$$f_t = \frac{v}{v^2/2a}$$

or

$$f_t = \frac{2a}{v}$$

as $v$ gets bigger $f_t$ gets smaller. Thus, the graph of $f_t$ is like that in Figure 27. For some value of $v$ the traffic flow reaches a maximum. In the next unit you will learn how to find this maximum value from the equation for the graph. For the present you can put some typical values into equation 29 to get an idea of the predictions of the model.

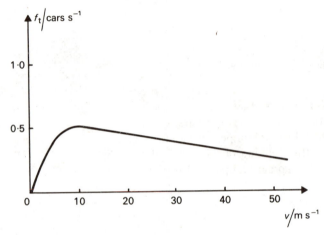

*Figure 27    A graph of traffic flow against velocity according to the third proposed model.*

SAQ 29

### SAQ 29

Taking $b = 4\,\text{m}$, $a = 6.5\,\text{m s}^{-1}$, $t_t = 0.6\,\text{s}$ fill in the gaps in the following table for $f_t$ and $v$ using equation 29

| $v/\text{m s}^{-1}$ | 0 | 5 | 10 | 20 | 30 | 40 |
|---|---|---|---|---|---|---|
| $f_t/\text{s}^{-1}$ | 0 | | | | | |

Hence check the curve of flow versus velocity predicted by this model and estimate the velocity which gives maximum traffic flow. Do you think this is realistic?

It turns out that in practice on single-lane roads the speed for maximum flow of traffic is about 15 mph or nearly $7\,\text{m s}^{-1}$, which suggests that this model is quite a good representation of reality at least as far as maximum flow is concerned. However, there is little doubt that few drivers in fact allow a distance as great as the stopping distance in front of the car in congested traffic conditions, but that they do allow a distance greater than the thinking distance.

SAQ 30

### SAQ 30

Try a different model. Suppose each driver allows twice the *thinking* distance plus half the *braking* distance between his own front bumper and the rear bumper of the car in front. Write an equation to express this model mathematically and estimate by drawing another curve on Figure 27 the speed of maximum flow.

Each model can be tested against reality by, for example, measuring average flow at different speeds. With a good model it is, for example, possible to build the right number of lanes to handle congested traffic at different speeds without severe hold ups.

## 5.3 How fast was the car going?

It is possible to reconstruct accidents of the type portrayed in Figure 28. If the car in Figure 28 were found in the sand on the beach a horizontal distance of 40 m from the edge of the cliff, and the cliff were 10 m high then it would be possible to calculate the speed at which the car was travelling as it went over the cliff edge.

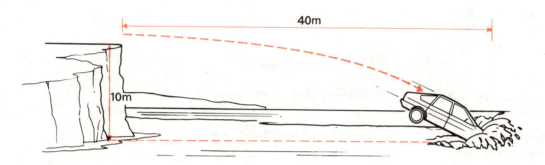

*Figure 28  The flight of a car falling at speed over a cliff edge.*

The simplest way to do this is to suppose that the car is a solid projectile, that air resistance is negligible and so use the model of a projectile described in Section 3.2. Let us also suppose that the car was travelling horizontally (i.e. in the $x$-direction) as it approached the cliff edge with a velocity $u_x$, so that the vertical component of velocity $u_y$ is zero. Let us regard the direction vertically *downwards* as the positive direction of $y$.

### SAQ 31

SAQ 31

Suppose the ground on top of the cliff was sloping downwards, towards the cliff edge, at an angle of 10° to the horizontal and the speed of the car was 20 m s$^{-1}$. What are the vertical and horizontal components of the car's velocity? (Figure 29.)

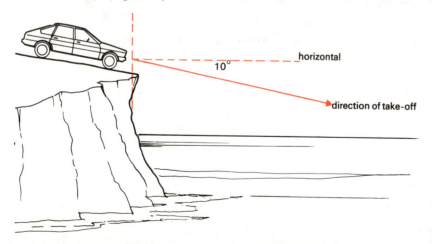

*Figure 29  Car leaving the cliff at an angle of 10° to the horizontal.*

If, however, we suppose there is no vertical component of velocity, $u_x$ is the quantity we want to calculate and $u_y$ is initially zero.

If $t_f$ is the time of flight of the car after it leaves the cliff and hits the sand, the distance it travels horizontally is $u_x t_f$.

So we know that $u_x t_f = 40\,\text{m}$ in this case.

Consider now the vertical component of velocity

| | |
|---|---|
| initial velocity | $u_y = 0$ |
| acceleration | $a = 9.81\,\text{m s}^{-2}$ |
| distance | $s = 10\,\text{m}$ |
| time | $t_f = \dfrac{40}{U_x}\,\text{s}\quad (U_x > 0)$ |

We want an equation relating these four quantities; namely $s = ut + \frac{1}{2}at^2$.

So
$$10 = \tfrac{1}{2}9.81 \times \frac{(40)^2}{(U_x)^2}$$

i.e.
$$10 = \frac{9.81 \times 1600}{2U_x^2}$$

So
$$U_x = \sqrt{\frac{9.81 \times 1600}{20}} = 28 \qquad \text{(to two significant figures)}.$$

So the horizontal speed of the car was initially about $28\,\text{m s}^{-1}$ or 63 mph according to this model.

How might the suppositions or assumptions used in this model affect the accuracy of the calculation?

At a speed of 63 mph it is doubtful whether the car would have stopped on hitting the sand; it would probably have bounced or skidded before coming to rest. It there are marks on the sand which are due to the car, closer to the cliff than 40 m, they should of course be used in the calculation. If any such marks have once been there but have now disappeared the calculated speed will be an over-estimate.

Air resistance, particularly if the car somersaulted, will have slowed the car down, so in that case the calculated speed will be an under-estimate. A strong wind blowing out to sea would however reduce this error.

The actual direction of take-off may be difficult to judge if the ground is bumpy. Small errors here can have big effects on the accuracy of the calculation.

*Example*

Assume that, as a result of a bump at the edge of the cliff, the car took off at an upward angle of 10° to the horizontal. What would its initial velocity have been in this case?

Here I must repeat the above calculation, and first solve for $u_x$ again, but this time I must include a value of $u_y$ derived as follows

$$u_y = -u \sin 10°$$

and $\quad u_x = u \cos 10°$

If I divide one equation by the other I get

$$\frac{u_y}{u_x} = \frac{-u \sin 10°}{u \cos 10°}$$

So $\quad u_y = -u_x \tan 10°$

Now consider the vertical component only.

We can proceed with the calculation as before. We know

that $\quad s_y = 10\,\text{m}$

$\qquad a = 9.81\,\text{m s}^{-2}$

$\qquad t_f = \dfrac{40}{U_x}\,\text{s}$

but this time $u_y = -u_x \tan 10°$

So $\qquad 10 = -U_x \tan 10° \dfrac{40}{U_x} + \frac{1}{2} \times 9.81 \times \dfrac{40^2}{U_x^2}$

therefore

$$10 + 40 \tan 10° = \dfrac{\frac{1}{2} \times 9.81 \times 1600}{U_x^2}$$

The tangent of 10° is 0.1763, so

$$U_x^2 = \dfrac{9.81 \times 1600}{2(10 + 7.05)} \qquad\qquad (30)$$

and $\qquad U_x = 21.5 \qquad$ (to three significant figures)

so $\qquad u_x = 21.5 \, \mathrm{m\,s^{-1}}$

This is the horizontal speed of the car. Its speed $u$ at an angle of 10° upwards is, as we noted earlier, found from

$$U_x = U \cos 10°$$

So $\qquad U = \dfrac{21.5}{\cos 10°} = 21.8$

This is only 49 mph as compared with over 60 mph for the horizontal take off.

**SAQ 32**

SAQ 32

What must the initial speed of the car have been if it left the cliff top at a *downward* angle of 10°?

### 5.4 The effects of a head-on crash in a car

One of the key factors affecting the safety of car passengers is the force they experience in a violent impact. Another factor is whether those in the front seats will hit the windscreen, or even go through it. Let us consider the dynamic factors in this case. We will use the example of a car crashing headlong into a solid wall.

(i) *The acceleration of the passenger compartment*

*Figure 30 The front of a car being compressed.*

First, I will consider the force on the passenger compartment as it slows down. If the front of the car is compressed by a length $d$ (Figure 30) and the car was travelling with initial velocity $u$ then I can use the equations in Section 3 to estimate the deceleration, the duration of the impact and the force on the passenger compartment.

In this calculation I make the following suppositions: (1) the car does not bounce back off the wall (2) it does not pitch or roll (3) the force acting on it is constant for the duration of the impact and (4) the mass of the part of the car which crumples is negligible compared with the total mass of the car.

The first two suppositions enable me to state that the final velocity after impact is zero.

The second two enable me to use Newton's Second Law with a constant force acting on a constant mass.

For any initial velocity $u$, the distance travelled by the passenger compartment is $d$, and I want to find the acceleration $a$ when the final velocity is zero. The equation which relates these quantities is

$$v^2 = u^2 + 2as$$

which becomes $0 = u^2 + 2ad$ in this case;

or  $\quad 2ad = -u^2$

and  $\quad a = \dfrac{-u^2}{2d}$  (31)

Thus, there is an average deceleration of $\dfrac{u^2}{2d}$ and the force producing it is given by Newton's Second Law $f = ma$.

If $m_c$ is the mass of the car, then the retarding force $f_c$ on the car is

$$f_c = \dfrac{-m_c u^2}{2d}$$

The force will thus be less if $d$ is made bigger. The more the front of the car is compressed the smaller the force experienced by the passenger compartment. If the length of the front of the car is doubled the force is halved. You will notice that the force $f_c$ increases with the square of the initial velocity and so if the speed doubles the force goes up by a factor of 4. If the wall, for example, which is hit in the collision cannot exert such a force then the wall will collapse and the car will not be completely stopped in the accident.

### SAQ 33

Using the same information as that used for deriving the acceleration of the passenger compartment, find an expression for the *time* it takes for the passenger compartment to come to rest.

(ii)  *The occupants and the windscreen*

If the occupants are firmly strapped to their seats by seat belts the deceleration of the passenger compartment will also be the deceleration of the passengers.

What force, therefore, must a strap exert on a passenger in order to hold him or her in place?

If we suppose a passenger stays quite rigid then the force is simply given by Newton's Second Law again. This time, however, the mass is equal to the mass of the passenger $m_p$.

So the force exerted by the seat belt is $\dfrac{m_p u^2}{2d}$

Suppose, however, that a passenger is not wearing a seat belt and that the restraining forces on him (or her) are negligible. The seat is smooth and the passenger is completely unprepared. Because there is no significant force acting upon the passenger the acceleration is zero and so the passenger continues to move with the initial velocity $u$ of the car. After a time $t$ the passenger will have moved forward a distance $x = ut$. The windscreen, however, will have an acceleration of $\dfrac{-u^2}{2d}$ (from equation 31) so I can find the distance it has moved at time $t$ using the formula

$$s = ut + \tfrac{1}{2}at^2$$

If the windscreen was initially a distance $w$ in front of the passenger's head, the distance $x$ that his head must travel before it hits the windscreen is

$$x = w + s$$

so

$$x = w + ut - \frac{u^2 t^2}{4d}$$

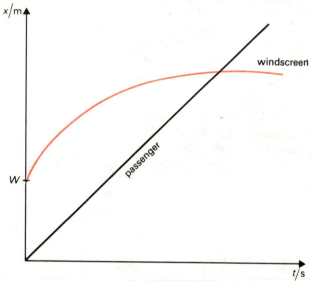

Figure 31   A graph of distance against time of passenger and windscreen.

Figure 31 shows a graph of the motion of both passenger and windscreen. You will see that in this case the passenger hits the windscreen. This will happen when $x$ and $t$ for the windscreen are the same as for the passenger. So we have two simultaneous equations

$$x = ut$$

$$x = w + ut - \frac{u^2 t^2}{4d}$$

which gives (by substituting $x = ut$ in the second equation)

$$\frac{u^2 t^2}{4d} = w$$

$$t = \frac{\sqrt{4dw}}{u}$$

This is the time which elapses before the passenger's head strikes the windscreen. The important factor is the *difference between the* passenger's velocity and the windscreen's velocity. The larger the difference the greater the injury and the greater the likelihood that the passenger's head will actually break the windscreen.

Now the passenger's velocity is $u$. How can I find the windscreen's velocity? I know the initial velocity $u$, that $a = \dfrac{-u^2}{2d}$ and that $t = \dfrac{\sqrt{4dw}}{u}$ and I want to find $v$, so I need to use

$$v = u + at$$

Thus,   $v = u - \dfrac{u^2}{2d}\dfrac{\sqrt{4dw}}{u}$

$$v = u - u\sqrt{\frac{w}{d}}$$

53

Now the difference in velocities (i.e. the passenger's velocity minus the windscreen's velocity) is the velocity of impact, given by

$$= u - \left( u - u\sqrt{\frac{w}{d}} \right)$$

So the velocity of impact is $= u\sqrt{\frac{w}{d}}$

The bigger $w$ (the initial distance between passenger and windscreen) the bigger this velocity and the more dangerous for the passenger. Hence, if a passenger is not wearing a seat belt this model indicates that it is safer for him or her to be as near the windscreen as possible. This is confirmed by findings of the Ministry of Transport. One reason why a *driver* without a seat belt often comes off better than a passenger without a seat belt is that the driver hits the steering wheel which is a shorter distance from him or her than is the windscreen from the passenger. Also, of course, impact with a steering wheel is not so dangerous.

Nowadays, seat belts are being made of extendable material to allow a passenger to move forward towards the windscreen on impact but not so far as to hit the windscreen. Why is this do you think? It is to reduce the force of the belt on the passenger—since his deceleration is slower—while still preventing his impact with the windscreen.

### 5.5 Impact with a crash barrier at the centre of a motorway

As you read in the road-safety essay one of the functions of a crash barrier is to redirect a vehicle that hits it back into the correct line of traffic. If the barrier was very strong and resilient a car might bounce back off it, again crossing the lines of traffic; so the barrier should give by just the right amount to reduce the vehicle's speed *at right angles to the barrier* to zero. At the same time it should not reduce its forward speed too much because this can cause somersaulting or spinning, etc. Let us see if we can use any of our mathematics to help us design a good crash barrier.

I would like you to try to set up an appropriate model, but I will prompt you with a series of questions, to which I have suggested solutions in the SAQ answers.

SAQ 34

**SAQ 34**

(a)  What is a reasonable forward speed to assume as typical for a vehicle leaving the fast lane of a motorway as a result of, for example, a tyre blow-out?

(b)  If a vehicle mounts the central reservation of the motorway at what angle might it strike the crash barrier?

(c)  If the barrier is 50 cm high by how much do you think it is reasonable to allow it to give sideways without breaking?

(d)  What is the simplest model that can be used to represent a car which is about to crash against the barrier?

(e)  What should the barrier be like if it is not to slow down the forward motion of the car?

(f)  What calculation would you perform to find the force at right angles to the motorway that the barrier must exert, and what further typical quantity must you estimate?

You should now be able to do the necessary calculations and be able to advise a design engineer what force his barrier should be able to exert.

# SUMMARY OF THE UNIT

*Newton's First Law* states that a body continues in a state of rest or uniform motion in a straight line unless acted upon by a force.

Section 2.1

*Newton's Second Law* states that the force exerted on a body is equal to the rate of change of its momentum. The *momentum* of a body is the product of its mass and its velocity. When the mass of a body is constant Newton's Second Law reduces to *force = mass × acceleration* or $f = ma$. The *mass* of an object does not change because it is a measure of the *amount of matter in it*. The *weight* of an object on Earth is the *force with which the Earth attracts it*. On the moon the weight would be the force with which the moon attracts it. *Newton's Law of Gravitation* states that two bodies of masses $m_1$ and $m_2$ and a distance $r$ apart attract each other with a force $\dfrac{Gm_1 m_2}{r^2}$ where $G$ is the gravitational constant. On the surface of the Earth the quantity $\dfrac{Gm_1}{r^2}$ is called the *acceleration due to gravity g*. On Earth the weight of a mass $m$ equals $mg$.

Sections 2.2, 2.3, and 2.4

$$G = 6.67 \times 10^{-11}\,\mathrm{m^3\,kg^{-1}\,s^{-2}} \quad \text{and} \quad g = 9.81\,\mathrm{m\,s^{-2}}$$

*Newton's Third Law* states that action and reaction are equal and opposite.

Section 2.5

Equations which relate the initial velocity $u$ final velocity $v$ and the distance travelled $s$ over a period of time $t$ for a constant acceleration $a$ are:

Sections 3.1 and 3.2.1

$$v = u + at \tag{11}$$
$$s = \tfrac{1}{2}(u + v)t \tag{12}$$
$$s = ut + \tfrac{1}{2}at^2 \tag{13}$$
$$v^2 = u^2 + 2as \tag{14}$$

and, less commonly used, $s = vt - \tfrac{1}{2}at^2$.

These equations can be used to model the motion of *projectiles*. In this case, there is a downward acceleration of $g$. The motion is *resolved into horizontal and vertical components* and *air resistance is neglected*. Under these conditions the angle of projections for *maximum range* on horizontal ground is 45° to the horizontal.

Section 3.2.2

A quadratic equation is an equation of the form $ax^2 + bx + c = 0$. $b$ and $c$ may be zero. The *solution* to this equation can be found by using the formula

Sections 4.1, 4.2, 4.3 and 4.4

$$x = \frac{-b \pm \sqrt{b^2 - 4ac}}{2a} \tag{23}$$

or it may be found by a graphical method, by completing the square or, in some cases, by factorization. (You are expected to be able to solve such equations either by *using the formula* or by *completing the square*.)

The square root of $-1$, $\sqrt{-1}$, is represented by the *imaginary number* i. i × i $= -1$. The square root of any negative number can be represented as an imaginary number. For example, $\sqrt{-16} = \sqrt{16} \times \sqrt{-1} = 4\mathrm{i}$. A *complex number* such as $A + B\mathrm{i}$ is the sum of a real number ($A$) and an imaginary number ($B\mathrm{i}$). The *modulus* of $A + B\mathrm{i}$ is $\sqrt{A^2 + B^2}$ and its *complex conjugate* is $A - B\mathrm{i}$. Quadratic equations with no real solutions have *complex solutions*—that is, the solutions are complex numbers.

Section 4.5

A parabola which opens downwards has a *maximum* value; one which opens upwards has a *minimum* value. For any parabola $Y = AX^2 + BX + C$ the maximum or minimum value will occur when $X = -B/2A$. If $A$ is *positive* the value of $Y$ reaches a *minimum*. If $A$ is *negative* then the value of $Y$ reaches a *maximum*.

*Maximization of tax revenue* is an example of a model which uses in the ideas of Section 4.

Section 4.7

Some cases where the ideas developed in the unit can be used in the

Section 5

mathematical representation of a model are in the context of *traffic and safety*. Examples are *skidding, traffic flow, reconstruction of accidents* where vehicles fall over a cliff or embankment and *head-on crashes*.

# ANSWERS TO SELF-ASSESSMENT QUESTIONS

## SAQ 1

(a) Force = mass × acceleration. For the Citroen both mass and acceleration are given, so

force (thrust) = 1320 × 0.65

= 858 N (to three significant figures).

(b) For the Volvo the acceleration is not given. However, the velocity increases by $10\,\mathrm{m\,s^{-1}}$ in 9.7 s.

so acceleration $= \dfrac{\text{increase in velocity}}{\text{time}}$

$= \dfrac{10}{9.7}\,\mathrm{m\,s^{-2}}$

Now you can calculate the force.

Force $= \dfrac{1280 \times 10}{9.7}$

= 1320 N (to three significant figures).

Suppositions:

(i) Acceleration is constant.

(ii) Thrust or force in top gear is constant.

(iii) There is no air resistance or retarding friction on the wheels.

(iv) The mass of the car is constant.

(v) The car travels in a straight line. (Further forces, not wholly at right angles to motion, come into play if the car is cornering.)

(c) If a car changes gear during acceleration the supposition that the force remains constant is clearly false. In lower gears at these speeds the thrust is increased as compared with top gear; which is why the acceleration of the Ford is so much greater than for the Volvo and the Citroen.

## SAQ 2

| Time/ number of intervals | Distance covered in corresponding interval/m | Total distance/m |
|---|---|---|
| 1 | 1 | 1 |
| 2 | 3 | 1 + 3 = 4 |
| 3 | 5 | 1 + 3 + 5 = 9 |
| 4 | 7 | 1 + 3 + 5 + 7 = 16 |
| 5 | 9 | = 25 |
| 6 | 11 | = 36 |
| $T$ | $2T - 1$ | $= T^2$ |

You are expected to notice, in doing this SAQ, that the numbers in the final column are the squares of the numbers in the first column. The expression $2T - 1$ in the middle column, which you were *not* asked to include in your answer, expresses the fact that each number in this column is one less than twice the number in the first column.

Notice that I have used the capital letter $T$ to denote the *number* of time intervals. That is $T$ is not a quantity with dimensions, it is just a number and I have therefore used a capital letter for it. The reason for this, is that the distance $T^2$ is a number of metres, it cannot have the dimensions of 'seconds squared'.

## SAQ 3

The weight of an object is the strength of the gravitational force upon it, which is given by

$\dfrac{Gm_1m_2}{r^2}$

Suppose the object has a mass $m_2$. The force upon it (i.e. its weight) will depend on the factor $m_1/r^2$. This quantity $m_1/r^2$ is less for an object on the surface of the moon than it is for the same object on the surface of the Earth, so it weighs less.

*Note:* To reveal the difference in weight you would have to use a spring balance, or 'feel' the weight. A pair of scales (i.e. a balance) using weights would not reveal the reduction in weight because the object and the weights would still balance.

## SAQ 4

You can write the first equation as

$\dfrac{\text{force on } m_2}{m_2} = \dfrac{Gm_1}{r^2}$

But the second equation tells you that

$\dfrac{\text{force on } m_2}{m_2} = g$ (i.e. $9.81\,\mathrm{m\,s^{-2}}$)

You can now substitute this into the first equation giving

$g = \dfrac{Gm_1}{r^2}$

You know all these quantities except $m_1$.

First rearrange the equation to make $m_1$ the subject of the equation, thus

$m_1 = \dfrac{gr^2}{G}$

Remembering that $r$ must be expressed in metres, $g$ is in metres per second squared ($\mathrm{m\,s^{-2}}$) and $G$ is also in SI units.

$m_1 = \dfrac{9.81 \times (6\,370\,000)^2}{6.67 \times 10^{-11}}\,\mathrm{kg}$

So $m_1 = 5.97 \times 10^{24}\,\mathrm{kg}$, which is the mass of the Earth.

## SAQ 5

The weight of a mass $m$ is the gravitational force exerted on it. On the surface of the moon this is given by

force on $m = \dfrac{G \times m \times \text{mass of the Moon}}{(\text{radius of the Moon})^2}$

If I put $\dfrac{G \times \text{mass of the Moon}}{(\text{radius of the Moon})^2} = g_{\text{moon}}$

where $g_{\text{moon}}$ is the acceleration due to gravity on the Moon, then

$g_{\text{moon}} = \dfrac{6.67 \times 10^{-11} \times 7.34 \times 10^{22}}{(1740 \times 10^3)^2}\,\mathrm{m\,s^{-2}}$

$= 1.62\,\mathrm{m\,s^{-2}}$

Compared with $g$ on Earth, $9.81\,\mathrm{m\,s^{-2}}$, the acceleration due to gravity on the Moon is about 1/6 as great. (Although the moon's mass is about 1/80th of the Earth's mass its radius is less by a factor of less than 4.)

## SAQ 6

The momentum of the bullet is given by the mass multiplied by the velocity. In SI units the mass is 0.020 kg and the velocity is $500\,\mathrm{m\,s^{-1}}$,

So momentum $= 500 \times 0.02\,\mathrm{kg\,m\,s^{-1}}$

$= 10\,\mathrm{kg\,m\,s^{-1}}$

This is the same as the momentum of the bullet and the mass after impact. The velocity of the two together is given by

momentum = (mass of bullet + mass of man) velocity

Rearranging gives

velocity $= \dfrac{\text{momentum}}{\text{total mass}}$

The momentum is the same as before impact so

$$\text{velocity} = \frac{10}{50.02}\,\mathrm{m\,s^{-1}} = 0.2\,\mathrm{m\,s^{-1}}$$

This is quite a slow speed (less than 1 mph) so a violent kick backwards would not be produced by this bullet.

## SAQ 7

To eliminate $t$ you should first rearrange equation 11 to make $t$ the subject.

Thus $v - u = at$ becomes $t = \dfrac{v - u}{a}$

Substituting this in equation 12 gives

$$s = \tfrac{1}{2}(u + v)\frac{(v - u)}{a}$$

so $\;s = \dfrac{1}{2a}(uv - u^2 + v^2 - uv)$

The two terms $uv$ have opposite signs so they subtract to zero

so $\;s = \dfrac{v^2 - u^2}{2a}$

Thus, multiplying both sides of the equation by $2a$ gives you

$$2as = v^2 - u^2$$

which can be written as

$$v^2 = u^2 + 2as$$

## SAQ 8

The fifth equation must omit the initial velocity $u$, because all the four equations 11–14 contain it. So it is necessary to substitute for $u$ in one equation, using its value from another one. You can do this with any pair of equations you care to choose.

From equation 11 $u = v - at$

(a) Substituting in equation 13 gives

$$s = (v - at)t + \tfrac{1}{2}at^2$$

so $\;s = vt - at^2 + \tfrac{1}{2}at^2$

so $\;s = vt - \tfrac{1}{2}at^2$; which is the required answer.

(b) Substituting in equation 14 gives

$$v^2 = (v - at)^2 + 2as$$

so, by multiplying out the squared bracket,

$$v^2 = (v^2 - 2avt + a^2t^2) + 2as$$

Subtracting $v^2$ from both sides gives

$$0 = -2avt + a^2t^2 + 2as$$

Rearranging

$$2as = 2avt - a^2t^2$$

Dividing through by $a$ and by 2 gives

$$s = vt - \tfrac{1}{2}at^2$$

## SAQ 9

(a) After the first interval of time the distance covered was 1 metre,

so $s = 1\,\mathrm{m}$, $t = 12\,\mathrm{s}$, $u = 0$

and you want to find $v$. The equation needed is

$$s = \frac{(u + v)}{2}t$$

Rearranging gives

$$v = \frac{2s}{t} - u$$

So $\;v = \dfrac{2 \times 1}{12}\,\mathrm{m\,s^{-1}} = 0.167\,\mathrm{m\,s^{-1}}$

After 24 s the distance covered is 4 m

So $\;v = \dfrac{2 \times 4}{24}\,\mathrm{m\,s^{-1}} = 0.333\,\mathrm{m\,s^{-1}}$

After 36 s the distance covered is 9 m

So $\;v = \dfrac{2 \times 9}{36}\,\mathrm{m\,s^{-1}} = 0.5\,\mathrm{m\,s^{-1}}$

And so on.

(b) The change of velocity in each interval of 12 s is $0.167\,\mathrm{m\,s^{-1}}$ or $1/6\,\mathrm{m\,s^{-1}}$

But acceleration $= \dfrac{\text{change of velocity}}{\text{time}}$

$$= \frac{1}{6 \times 12}\,\mathrm{m\,s^{-2}} = \frac{1}{72}\,\mathrm{m\,s^{-2}}$$

$$= 0.0139\,\mathrm{m\,s^{-2}}$$

## SAQ 10

You know that $D_b = U^2/13$ so you can complete the table as follows

| $U$ | 0 | 5 | 10 | 20 | 30 | 40 | 50 |
|-----|---|-----|-----|------|------|-----|-----|
| $D_b$ | 0 | 1.9 | 7.7 | 30.8 | 69.2 | 123 | 192 |

Your graph should be as in Figure 32.

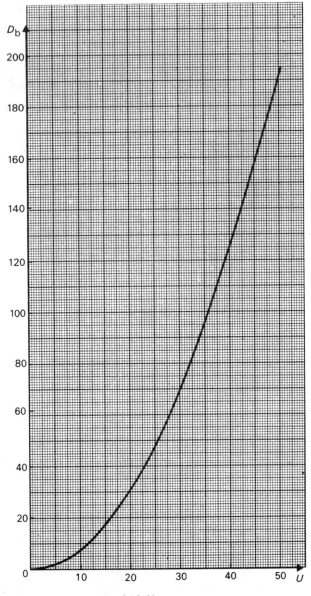

*Figure 32   Answer to SAQ 10.*

## SAQ 11

Substituting 20, 30 and 40 for $U$ in equation 18 gives the following values for $D_s$:

42.8, 87.2, 147.

These are numbers of metres.

The distances given in Figure 10 are shown as multiples of 5 metres or 10 metres.

## SAQ 12

Substitute the number 40/2.25 for $U$ into equation 19 to obtain a value for $D$. This gives 48.6. So the sign should be at 50 or 60 metres away. (Alternatively, you could use Figure 14.)

## SAQ 13

If the initial velocity is $10\,\mathrm{m\,s^{-1}}$ then

$$s_y = u^2/2g = S_y\,\mathrm{m}$$
$$S_y = 100/9.81 \times 2$$
$$= 5.10 \quad \text{(to three significant figures)}$$

so the distance is 5.10 m.

## SAQ 14

The dimensions of velocity $u_y$ are distance/time, while those of $g$, the acceleration, are $\dfrac{\text{distance}}{\text{time}^2}$.

So the dimension of $\dfrac{u}{g}$ are $\dfrac{\text{distance}}{\text{time}} \times \dfrac{\text{time}^2}{\text{distance}}$

This reduces to the dimension of *time*.

In the problem set you know that $u = 10\,\mathrm{m\,s^{-1}}$, $v = 0$, $a = -9.81\,\mathrm{m\,s^{-2}}$ and you have to find $t$; so use $v = u + at$, with $t = T\,\mathrm{s}$.

Thus, $0 = 10 - 9.81T$

So $\quad T = \dfrac{10}{9.81} = 1.02$

and $\quad t = 1.02\,\mathrm{s}$

## SAQ 15

The time the ball takes to return to its starting point is just twice the time it takes to reach the top of its flight. This time is given by

$$v = u + at$$

where $\quad v = 0 \quad$ and $\quad a = -g$

So $\quad t = \dfrac{u}{g}$

Thus, the total time to return to the starting point is $2u/g$.

## SAQ 16

This time both $u_x$ and $u_y$ are the same because $\sin 45° = \cos 45° = 0.7071$. So for vertical flight, $u_y = 70.7\,\mathrm{m\,s^{-1}}$ with the projectile finishing where it started.

Using $s_y = u_y t - \frac{1}{2}at^2$ with $a = -9.81\,\mathrm{m\,s^{-2}}$ and $t = T\,\mathrm{s}$,

$$0 = 70.7T - \tfrac{1}{2}9.81T^2$$

or $\quad T\left(T - \dfrac{70.7 \times 2}{9.81}\right) = 0$

and $\quad T = 0$ or 14.4

Hence $\quad S_x = U_x T = 70.7 \times 14.4$

$$= 1020 \quad \text{(to three significant figures)}$$

The projectile will travel a horizontal distance of 1020 m.

The highest point it reaches is found by looking at vertical motion and setting $v = 0$, given $u_y = 70.7\,\mathrm{m\,s^{-1}}$ and $a = -g$. So use equation 14 which gives $s = 255\,\mathrm{m}$.

The projectile rises to a height of 255 m.

## SAQ 17

In the worked example $\dfrac{S_x}{U_x}$ would be written as

$$\frac{S_x}{U_x} = \frac{883}{86.6} = 10.2$$

If this is substituted into equation 22, it should give a value of $S_y = 0$:

Thus, $\quad S_y = 50 \times 10.2 - \dfrac{9.81}{2}(10.2)^2$

$$= 510 - 510.3$$
$$= 0 \quad \text{(within the accuracy of the calculations)}$$

Similarly, for your answer to SAQ 16

$$S_y = 70.7 \times \frac{1020}{70.7} - \frac{9.81}{2}\frac{(1020)^2}{(70.7)}$$

$$= 1020 - 1021$$
$$= 0 \quad \text{(to three significant figures)}$$

So solving equation 22 would have given the same results.

## SAQ 18

The graph is shown in Figure 33.

(a) $X = 0$ or $-4$.

(b) $X^2 + 4X - 3 = 0$ may be written
$$X^2 + 4X = 3$$
and $X^2 + 4X$ equals 3 at $-4.6$ or 0.6.

(c) $X = -2$.

(d) No solution since $X^2 + 4X$ does not go as low as $-5$.

## SAQ 19

If $\quad (X + 2)^2 = 7$,

$$X + 2 = \pm\sqrt{7}$$

So $\quad X = -2 \pm \sqrt{7}$

that is $\quad X = -2 + \sqrt{7}$ or $-2 - \sqrt{7}$

$$\sqrt{7} = 2.65 \quad \text{(to three significant figures)}$$

so $\quad X = 0.65$ or $-4.65$.

## SAQ 20

(a) $X^2 + 2X + 1$ can be written as
$$(X + 1)^2 = 0$$
So $X + 1 = 0$
So in this case $X = -1$ is the only (double) solution.

(b) $X^2 + 2X = 0$ can be written as
$$(X + 1)^2 - 1 = 0$$
So $(X + 1)^2 = 1$

You now take the square root of both sides
So $X + 1 = \pm 1$
or $X = 0$ or $-2$

You can also solve this equation by first noting that it can be written

$$X(X + 2) = 0$$

from which it can be seen at once that

$$X = 0 \text{ or } X = -2$$

for the equation to be satisfied.

59

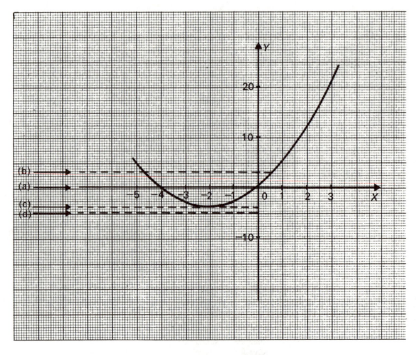

*Figure 33    Answer to SAQ 18. The arrows show the values to which the problem refers.*

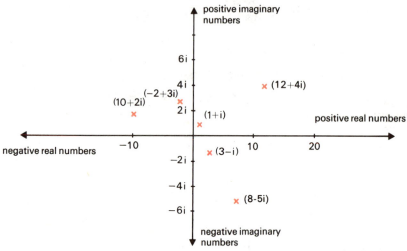

*Figure 34    Answer to SAQ 23.*

(c)    $X^2 - 2X - 3 = 0$ can be written as

$(X - 1)^2 - 4 = 0$

So $(X - 1)^2 = 4$

and $X - 1 = \pm 2$

Hence $X = 3$ or $-1$

(d)    $3X^2 - 12X - 1 = 0$   can be written as

$3(X^2 - 4X) - 1 = 0$

but $3(X^2 - 4X)$ can be written as
$3[(X - 2)^2 - 4] = 3(X - 2)^2 - 12$

So $3(X - 2)^2 = 13$

or $(X - 2)^2 = \dfrac{13}{3}$

Therefore, $X - 2 = \pm\sqrt{\dfrac{13}{3}}$

and $X = 2 \pm \sqrt{\dfrac{13}{3}}$

But $\sqrt{\dfrac{13}{3}} = 2.08$   (to three significant figures)

So $X = 4.08$ or $-0.08$.

**SAQ 21**

(a)    $X^2 + 6X = 0$

In this case, $A = 1$, $B = 6$ and $C = 0$ so

$X = \dfrac{-6 \pm \sqrt{6^2}}{2}$

$= \dfrac{-6 + 6}{2}$   or   $\dfrac{-6 - 6}{2}$

$= 0$   or   $-6$.

You can see this answer is correct from the factors $X(X + 6) = 0$.

(b)    $9X^2 + 12X + 2 = 0$

In this case $A = 9$, $B = 12$, $C = 2$ so

$X = \dfrac{-12 \pm \sqrt{12^2 - 4 \times 9 \times 2}}{18}$

$= \dfrac{-12 \pm \sqrt{144 - 72}}{18}$

$= \dfrac{-12 \pm \sqrt{72}}{18}$

$= \dfrac{-12 \pm \sqrt{36\sqrt{2}}}{18}$

60

Divide top and bottom by 6:

$$= \frac{-2 \pm \sqrt{2}}{3}$$

So  $x = \dfrac{-2 - \sqrt{2}}{3}$  or  $\dfrac{-2 + \sqrt{2}}{3}$

$$= -1.14 \quad \text{or} \quad -0.195$$

(c)  If $3X^2 - 6X + 10 = 0$, $A = 3$, $B = -6$, $C = 10$.

So    $X = \dfrac{+6 \pm \sqrt{6^2 - 4 \times 3 \times 10}}{6}$

$$= \frac{6 \pm \sqrt{-84}}{6}$$

There is no normal 'real' number which is the square root of $-84$ so there are no normal solutions to this equation.

## SAQ 22

(a)  In $X^2 + 2X + 2 = 0$, $A = 1$, $B = 2$, $C = 2$.

So $X = \dfrac{-2 \pm \sqrt{4 - 4 \times 2}}{2}$

$$= \frac{-2 \pm \sqrt{-4}}{2}$$

$$= \frac{-2 \pm 2i}{2}$$

So $X = -1 + i$  or  $-1 - i$.

(b)  In $X^2 - 4X + 7 = 0$, $A = 1$, $B = -4$, $C = 7$.

So $X = \dfrac{4 \pm \sqrt{16 - 4 \times 7}}{2}$

$$= \frac{4 \pm \sqrt{-12}}{2}$$

$$= \frac{4 \pm \sqrt{4}\sqrt{3}\sqrt{-1}}{2}$$

Dividing top and bottom by $2 = \sqrt{4}$ gives

$$= 2 \pm \sqrt{3}i$$

So $X = 2 + \sqrt{3}i$  or  $2 - \sqrt{3}i$

(c)  In $X^2 + 5X + 3 = 0$, $A = 1$, $B = 5$ and $C = 3$.

So $X = \dfrac{-5 \pm \sqrt{25 - 12}}{2}$

$$= \frac{-5 \pm \sqrt{13}}{2}$$

This has *real* roots

and   $X = \dfrac{-5 + \sqrt{13}}{2}$  or  $\dfrac{-5 - \sqrt{13}}{2}$

$$= -0.697 \quad \text{or} \quad -4.30 \quad \text{(to three significant figures)}$$

## SAQ 23

Figure 34 shows the numbers represented on the Argand diagram.

## SAQ 24

(a)  The moduli of the given complex numbers are

(i)  $\sqrt{5^2 + 12^2} = \sqrt{25 + 144} = \sqrt{169} = 13$

(ii)  $\sqrt{2^2 + 3^2} = \sqrt{4 + 9} = \sqrt{13} = 3.606$

(iii)  $\sqrt{A^2 + B^2}$

(b)  The complex conjugates are (i) $5 - 12i$, (ii) $2 + 3i$ (iii) $-A + Bi$.

The moduli of the complex conjugates are the same as the moduli of the original complex number. This is because the modulus (or magnitude) is not changed just because the vector joining the complex number to the origin is moved to a different quadrant of the diagram by changing a sign.

## SAQ 25

Since the coefficient of $X^2$ is negative the parabola $Y = 1 + 4X - 3X^2$ will have a maximum value.

It occurs when $X = \dfrac{-4}{2(-3)} = \dfrac{2}{3}$

and at this point $Y = 1 + 4(\tfrac{2}{3}) - 3(\tfrac{2}{3})^2 = 2\tfrac{1}{3}$

## SAQ 26

To solve these equations for $\bar{p}$ you must eliminate $\bar{q}$ between the equations. In this case, this is easy because each equation has $\bar{q}$ as the subject of the equations. Thus

$$c - d\bar{p} = a + b(\bar{p} - t).$$

Collecting terms in $\bar{p}$ on the right gives

$$c - a + bt = b\bar{p} + d\bar{p}$$

that is

$$\bar{p}(b + d) = c - a + bt$$

and

$$\bar{p} = \frac{c - a + bt}{b + d}$$

## SAQ 27

Comparing the equations given with the equations 24 and 25, and using capital letters because we are now working in numbers, you can see that

$$C = 400\,000$$
$$D = 1500$$
$$A = 100\,000$$
$$B = 8300$$

Using the result obtained in equation 26

$$T = \frac{8300 \times 400\,000 + 1500 \times 100\,000}{2 \times 8300 \times 1500}$$

$$= \frac{33.2 \times 10^8 + 1.5 \times 10^8}{24.9 \times 10^6}$$

$$= \frac{34.7}{24.9} \times 10^2$$

$$= 139$$

So the tax per gallon for maximum revenue is £1.39!

## SAQ 28

For this problem you use the fact that the speeds of the two vehicles are proportional to the square roots of their skid lengths.

Here, for the test vehicle, $u_t = 10\,\mathrm{m\,s}^{-1}$

$$s_t = 20\,\mathrm{m}$$

The distance $s$ for the crashed car is $72\,\mathrm{m}$. So since

$$\frac{u}{u_t} = \sqrt{\frac{s}{s_t}}$$

$$U = 10\sqrt{\frac{72}{20}}$$

$$= 10\sqrt{3.6} = 19 \quad \text{(to two significant figures)}$$

and so  $u = 19\,\mathrm{m\,s}^{-1}$

So the speed of the car was just under $19\,\mathrm{m\,s}^{-1}$ or about 42 mph, and the driver was exceeding the speed limit. If the road conditions were the same as when the accident occurred—particularly in respect of the wetness of the road—the estimate should be enough to condemn the car driver. The car's speed could be more than 42 mph because his car may not have stopped at the end of the skid; so his speed was *at least* 42 mph.

## SAQ 29

The table to be compiled is obtained by substituting values in the equation

$$f_t = \frac{v}{b + \dfrac{v^2}{2a} + vt_t}$$

Thus

| $v/\mathrm{m\,s^{-1}}$ | 0 | 5 | 10 | 20 | 30 | 40 |
|---|---|---|---|---|---|---|
| $f_t/\mathrm{s^{-1}}$ | 0 | 0.560 | 0.565 | 0.428 | 0.328 | 0.265 |

The speed for maximum flow rate according to this model is about $7\,\mathrm{m\,s^{-1}}$.

## SAQ 30

This new model gives rise to a modified expression. The distance between the cars will be

$$b + \frac{v^2}{4a} + 2vt_t$$

since $\dfrac{v^2}{2a}$ is the braking distance, which is to be halved, and $vt_t$ is the thinking distance, which is to be doubled. The new expression for flow becomes

$$f_t = v/(b + 2vt_t + v^2/4a)$$

Calculating the flow values according to this model gives

| $v/\mathrm{m\,s^{-1}}$ | 0 | 5 | 10 | 20 | 30 | 40 |
|---|---|---|---|---|---|---|
| $f_t/\mathrm{s^{-1}}$ | 0 | 0.456 | 0.504 | 0.461 | 0.402 | 0.352 |

The speed of maximum flow is now about $10\,\mathrm{m\,s^{-1}}$, as shown in Figure 35.

## SAQ 31

The vertical component of velocity is $20\sin 10°\,\mathrm{m\,s^{-1}}$ downwards, and the horizontal component of velocity is $20\cos 10°\,\mathrm{m\,s^{-1}}$. So

$$u_y = +3.47\,\mathrm{m\,s^{-1}} \quad \text{(to three significant figures)}$$

$$u_x = +19.7\,\mathrm{m\,s^{-1}} \quad \text{(to three significant figures)}$$

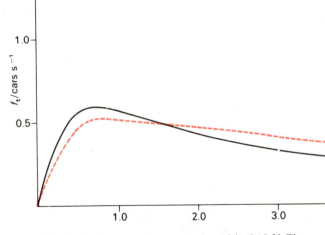

Figure 35   *The broken curve is the curve for the model in SAQ 30. The solid curve is a repeat of Figure 27.*

## SAQ 32

The difference between this question and the worked example is that now the vertical component of the initial velocity is downwards and is therefore positive. Thus

$$u_y = u_x \tan 10.$$

So equation 30 becomes

$$U_x^2 = \frac{9.81 \times 1600}{2(10 - 7.05)}$$

so   $U_x = 51.6$   (to three significant figures)

and   $u_x = 51.6\,\mathrm{m\,s^{-1}}$

This corresponds to a velocity of 115 mph. I would be inclined to doubt the validity of this result unless the cliff was at the bottom of a long hill. I would be inclined to think that the car had hit the ground nearer to the cliff and had bounced on, rather than accept that such a speed is true.

## SAQ 33

The initial velocity is $u$, the final velocity is 0, the distance travelled is $d$ and it is required to find $t$. The equation required is

$$s = \frac{(u + v)t}{2}$$

and   $d = \dfrac{(u + 0)t}{2}$

so   $t = \dfrac{2d}{u}$

## SAQ 34

(a)   About 60 mph or $27\,\mathrm{m\,s^{-1}}$. Even though the vehicle was in the fast lane it is likely to have managed to slow down a little before crashing.

(b)   Say 10°.

(c)   Perhaps 30 cm.

(d)   As a solid body which will simply bump into the barrier.

(e)   It should be smooth and present no impediment to forward movement.

(f)   Use Newton's Second Law, $f = ma$. You will have to estimate what mass of a vehicle to insert into the equation. An over-estimate is wise since a vehicle breaking through the barrier is more serious than one that bounces off. Perhaps 5 tonnes or 5000 kg.

*Calculation*

The component of velocity of the vehicle at right angles to the barrier is $27 \times \sin 10° = 4.7\,\mathrm{m\,s^{-1}}$.

This velocity must be reduced to zero in a distance of 30 cm or 0.3 m. So the acceleration is given by equation 14:

$$0 = 4.7^2 + 2A \times 0.3$$

So   $a = -37\,\mathrm{m\,s^{-2}}$   (to two significant figures)

Putting this in the equation $f = ma$ gives

$$f = -5000 \times 37\,\mathrm{N}$$

$$= -190\,000\,\mathrm{N} \quad \text{(to two significant figures)}$$

This is the force the barrier should exert as it is bent backwards a distance of 30 cm.

**Modelling by Mathematics**